And others, Joseph Gay-Lussac, J. S. Ames

The Free Expansion of Gases

.

And others, Joseph Gay-Lussac, J. S. Ames

The Free Expansion of Gases

ISBN/EAN: 9783337277536

Printed in Europe, USA, Canada, Australia, Japan

Cover: Foto ©berggeist007 / pixelio.de

More available books at **www.hansebooks.com**

THE

FREE EXPANSION OF GASES

MEMOIRS BY GAY-LUSSAC, JOULE
AND JOULE AND THOMSON

TRANSLATED AND EDITED

By J. S. AMES, Ph.D

PROFESSOR OF PHYSICS IN JOHNS HOPKINS UNIVERSITY

NEW YORK AND LONDON

HARPER & BROTHERS PUBLISHERS

1898

PREFACE

THE experiments on the changes of temperature which gases experience when they are allowed to expand under such conditions as to do no external work are of great importance from two standpoints : 1. Owing to the minuteness of the change of temperature, it may be assumed that the intrinsic energy of a gas is almost entirely kinetic ; and so conclusions may be drawn between the mechanical work done in compressing a gas and the rise of temperature produced : 2. Measurements of the minute changes lead to a method of comparing temperatures as registered on a gas-thermometer and those which are given on Thomson's Absolute Scale.

Robert Mayer assumed that there was no "internal work" done in compressing a gas, and so made a calculation for what is now called the mechanical equivalent of heat. The question as to whether Mayer was acquainted with the experiments of Gay-Lussac at the time he made his calculation or not has long been an open one ; but it is generally acknowledged now that he was familiar with the results obtained by Gay-Lussac, and so was justified in his theoretical work. Gay-Lussac performed his experiments on the free expansion of gases in the year 1807 ; and a translation of his memoir is given in this volume.

In 1844 Joule repeated these experiments, being unacquainted with the work of Gay-Lussac. His published account of his research follows. The important bearing of Joule's experiment upon the science of Thermodynamics was recognized by William Thomson (now Lord Kelvin) ; and he began, in collaboration with Joule, a more complete investigation of the "Thermal Effects of Fluids in Motion," which lasted for many years. The most important of their experiments are reported in this volume. Thomson applied his dynamical theory of heat to the results obtained, and thus secured comparisons between gas-thermometer temperatures and those on an "absolute" scale, one of the most important contributions of the century to physical science.

GENERAL CONTENTS

First Attempt to Determine the Changes in Temperature which Gases Experience owing to Changes of Density, and Considerations on their Capacity for Heat.

BY

L. J. GAY-LUSSAC, Mémoires d'Arcueil, **I.** 1807.

Reprinted in *Die Principien der Wärmelehre*.
E. MACH. Leipzig, 1896.

CONTENTS

First Attempt to Determine the Changes in Temperature which Gases Experience owing to Changes of Density, and Considerations on their Capacity for Heat.

GAY-LUSSAC.

Read at the Institute, September 15, 1806.

In the researches published by Mr. Humboldt and myself on eudiometric methods and the analysis of atmospheric air, we have noticed that the combustion produced by an electric spark in a mixture of oxygen and hydrogen was not complete when the two gases were in the ratio of 10 to 1. In this experiment, when the excess of oxygen over and above that necessary for the saturation of the hydrogen was replaced by nitrogen, the combustion stopped at almost exactly the same point as before. Guided by special considerations, we were led to think that this phenomenon depended on the fact that, since the heat set free in the combination was absorbed by the parts of each gas which had not entered into combination, the temperature was lowered below the point necessary for combustion, and, as a consequence, combustion ceased. Since, further, we saw that the action of nitrogen in this respect was almost identical with that of oxygen, we had assumed that the reason why these two gases stopped the combustion at the same point was because they certainly had the same capacity for heat. We were not able to verify our conjectures in the case of other gases; but, as one is naturally inclined to generalize, we maintained the opinion—I in particular—that it was most probable that all gases had the same capacity for heat. On my return to Paris, after a journey which I made with Mr. Humboldt to Italy and Germany, I was impatient to make some more direct experiments in order to see to what extent our first conjectures were well founded, being convinced that, whatever was the result, I would not have

3

worked in vain. I communicated my project to Mr. Berthollet, who encouraged me to execute it ; and he, as well as Mr. Laplace, have taken the most lively interest in it. If it is flattering to me to be able to mention here the names of these two illustrious scientists who honor me with their esteem, it is my duty to state at the same time that I owe a great deal to their clear-sighted advice. It was at Arcueil, in the laboratory of Mr. Berthollet, that my experiments were made. They have led me, so far as the capacity of gases is concerned, to results quite unexpected and contrary to those which I had suspected, and have attracted my attention to several new phenomena which appear to have a most important bearing upon the theory of heat.

Starting from the two facts—that all gases are expanded equally by heat, and that they occupy volumes which are inversely proportional to the weights which compress them—I thought, with Mr. Dalton,* that by putting them all under the same conditions and then decreasing by the same amount the pressure common to them all, it would be possible to see from the changes of temperature produced by the increase in volume whether they had or had not the same capacities for heat. It was to this end that I used the following apparatus :

I took two balloon flasks, each having two openings and each of twelve litres' capacity. Into one of the openings of each flask I fitted a cock, and into the other a very sensitive alcohol thermometer, whose centigrade degrees could easily be read to hundredths. I at first used an air thermometer, constructed on the principles of Count Rumford or Mr. Leslie ; but although infinitely more sensitive than the alcohol thermometer, it was, in several respects, inconvenient. I can remedy these defects now ; but they made me prefer the alcohol thermometer, because it gave me results which were more comparable among themselves. In order to avoid effects due to moisture, I introduced dried calcium chloride into each flask. The arrangement of the apparatus for each experiment was as follows : Both flasks being exhausted, and having assured myself that there was no leakage, I filled one flask with the gas upon which I wished to experiment. About twelve hours later I connected the two flasks by a lead tube, and, opening the cocks, the gas

* *Journal des Mines*, tom. 13, p. 257.

rushed into the empty flask until equilibrium of pressure was established. During this time the thermometer experienced changes which I carefully noted.

I began my experiments with this apparatus by using atmospheric air; and I observed, with MM. Laplace and Berthollet, that the air on passing into the empty flask from the other made the thermometer [*in the former*] rise, as several physicists have previously noted. It was known that when air expands, owing to a decrease in pressure, it absorbs heat; and, *vice versa*, that on being compressed it sets heat free. From this fact several physicists had drawn the conclusion that the capacity for heat of air which is expanded is greater than that of compressed air, and that an empty space should contain more heat than the same space filled with air. Considering equal weights of this fluid under different pressures and at the same temperature, there is no doubt but that the more it is expanded the more heat it contains, because as it expands it absorbs continually. But when we consider equal volumes, nothing justifies us in believing that the same thing must be true. If in our experiment the expanded air which remains in the flask originally filled has absorbed heat, that which has left the flask has carried heat with it, and it is not proved that the quantity of heat absorbed is greater than that taken away. Consequently, the opinion of those who believe that a vacuum contains more heat than a space full of air, an opinion which rests on the above considerations alone, is absolutely unfounded. We cannot believe, with Mr. Leslie, that it is the trace of air left in the receiver in virtue of the imperfect vacuum which gives rise to all this heat, owing to the great reduction of volume which it experiences, as a consequence of the air being admitted. If this were so, it would necessarily happen that, on introducing a very small volume into an absolutely empty receiver, there would be a quantity of heat absorbed nearly equal to that set free when the receiver is emptied of air to this same extent, and we let it fill completely. But, far from this, there is heat set free in every case. It may seem indifferent at first sight whether it is from a space which is empty or from one filled with air in a state of great expansion that the heat is liberated when air enters into this space; but it seems to me that for the theory of heat it is of the greatest importance to know the source of the heat. In my experiments, in spite of the most perfect vacuum which I could produce in my receiver, I have always seen the thermometer placed in it rise to a most marked degree when air from the other rushed into it; and I cannot avoid the conclusion that the heat does not come from the traces of air which may possibly be present.

Being convinced of this important fact, that the higher the vacuum in a receiver so much the greater is the amount of heat set free when the exterior air enters, I sought to determine by exact experiments what relation there was between the heat absorbed in one of the receivers and that set free in the other, and how these changes in temperature depended upon the differences in density of the air. For brevity's sake I shall call " No. 1 " the flask in which is enclosed the gas which is made the subject of experiment; and " No. 2," that which is empty. It is in the first that cold is produced; in the second, heat. In each experiment I have noted exactly the thermometer outside and the barometer; but, as one varied only between 19° and 21° C., and the other between $0^m.755$ and $0^m.765$, the corrections which should be made to the results are quite small, and can be neglected. In order to see what connection there was between the densities of the air and the changes of temperature which are due to differences of density, I used in succession air whose density decreased as the numbers 1, $\frac{1}{2}$, $\frac{1}{4}$, etc. In order to do this, after having made air pass from receiver No. 1 into the empty receiver No. 2, I renewed the vacuum in the latter, and waited until there was complete equilibrium of temperature between them both. Since the two receivers had equal volumes, the density of the air was thus reduced one-half. On opening the cocks, the air was again divided between the two flasks, and the density was reduced to one-quarter. I could have carried the reduction in a similar manner to $\frac{1}{8}$, $\frac{1}{16}$, etc.; but I stopped at $\frac{1}{4}$, because below that the changes in temperature, which continued to diminish, could have been observed accurately only with the greatest difficulty. The following table contains the means of the results of six experiments which I made on atmospheric air:

Density of Air Expressed by the Barometer.	Cold Produced in Flask No. 1.	Heat Produced in Flask No. 2.
$0^m.76$	$0°.61$	$0°.58$
$0^m.38$	$0°.34$	$0°.34$
$0^m.19$	$0°.20$	$0°.20$

In this table I give the records of only the means of the results, because the greatest variation above or below this mean have been but 0.05, when the density of the air was that expressed by $0^m.76$; and they were much smaller when the densities were those expressed by $0^m.38$ and $0^m.19$.

On comparing the results, it is seen that the heat absorbed by the air of flask No. 1 in the first experiment is $0°.61$, while that liberated in receiver No. 2 is only $0°.58$. The difference between these numbers is of itself sufficiently small to be attributed to some circumstance whose influence one might overlook, or even to errors of observation; but, if we consider the results given in the second and third rows, we see that the temperature changes are exactly equal to each other. I think, therefore, that I am justified in concluding that, when a given volume of air is made to pass from one receiver into another which is empty and of the same volume, the temperature changes in each receiver are the same.

The numbers 0.61, 0.34, and 0.20, which express these changes of temperature, are not exactly in the same ratio as the densities of the air; they diminish according to a law which is less rapid. But if we consider that in each experiment the time required in order that the entire effect should be produced was about two minutes, and that in equal intervals of time the coolings and heatings [*due to external causes*] increase with the difference in temperature between the media, we understand why the number 0.20 is more removed from the fourth of 0.61 than 0.34 is from the half. And if we are willing to admit this cause as that which produces these differences, we conclude that it is probable that, when we condense or expand air, the changes in temperature which it experiences are proportional to the differences of density.

If, then, the number 0.20 has been less affected by the sources of error than the others, it must be more exact than they; and, consequently, in accordance with the ratio just established, the number 0.61, which expresses the change in temperature of air when its density is $0^m.76$, is too small—it should be at least 0.80.

Nevertheless, this last number does not yet express exactly all the heat which has been absorbed or set free. To gain an idea of this quantity, it would be necessary to take into account the masses of the receivers and the thermometer, which

7

are considerable in comparison with the mass of the air. An air thermometer placed in the same conditions as the alcohol one indicated 5°.0 instead of 0°.61, as given by the latter.

As I must return later to this subject, after performing experiments which will be specially directed to this end, I shall say no more about the matter now. I shall note merely that the heat liberated or absorbed is very large compared with the mass of the air.

In order to avoid the effects of moisture, I was obliged to use two receivers, in one of which was placed calcium chloride. But when I made the exterior air enter directly into the empty receiver, the thermometric effects were nearly doubled—a fact in accord with the law which we have just established.

This law, that the thermometric effects follow the same ratio as the densities of the air, leads us to conclude that on suddenly diminishing or increasing a perfectly empty space no change of temperature would be produced. I have diminished the vacuum space of a large barometer in which I placed one of the bulbs of a sensitive air thermometer, and neither by inclining the barometer nor by then raising it erect did I perceive a change of temperature.

After these experiments it was extremely interesting to know what would happen with hydrogen, whose specific gravity is so different from that of atmospheric air. I filled receiver No. 1 with this gas, and, after leaving it twelve hours in contact with calcium chloride, during which time I took care to allow from time to time gas to enter, so as to fill the space left by the moisture as it was absorbed, I opened the communication with the empty receiver No. 2. The flow of the hydrogen was instantaneous compared with that of air, and the changes of temperature were much greater. The opening of communication between the two receivers had remained the same for the two gases ; and considering the great difference between their specific gravities, it was not difficult to recognize in this the true cause of the inequality of the times of escape. When, in fact, two fluids compressed to the same extent escape by two small openings of the same size, their velocities vary inversely as the square root of their densities. If, then, it is desired in our experiments to make the times of escape the same, the openings must be made inversely proportional to the square roots of the densities.

Making use of these principles, Mr. Leslie has devised a most elegant method for measuring the specific gravities of elastic fluids. Let us imagine a bag full of gas, and communicating by means of a cock which has a small opening with a bell-jar full of water and inverted over a large bath of the same liquid. When the cock is opened the gas passes out of the bag into the receiver, because there is no longer equilibrium of pressure, and a certain time is required for it to depress the water and make it come down to a definite point. On noting the time required by each gas to make the water reach the same mark, the specific gravities will vary directly as the squares of these times.*

In order to compare the effects of different gases with reference to the changes in temperature which they can produce in changing volume, it was necessary to make the conditions the same for them all, and, consequently, to modify my apparatus. First of all, it was necessary to have a means of measuring the time of escape for a given opening, and then to have a means of varying the openings in order to make the time of escape constant.

To satisfy the first requirement I placed a small disk of paper, 2 cm. in diameter, under the opening of the cock of the empty flask. This disk is supported by a ring of iron wire, carrying a small prolongation to serve as a lever and to bear a counterpoise. Two silk threads serve as the axis of the lever, and tend by a slight torsion, which can be given them, to bring the disk back into a horizontal position, in which a stop prevents it from going farther in that direction. When a gas enters the flask it strikes the disk, makes it take a vertical position, in which there is a second stop preventing further motion, and the time taken for the escape of the gas is measured by that taken by the disk to return to its horizontal position.

In order to vary the size of the opening at will I had Mr. Fortin construct for me a small piece of apparatus, which I shall briefly describe. It is a metallic disk in which there is an opening bounded by two concentric circles and by two radii, making an angle slightly less than 180°. A second disk, which is semi-circular, turns with friction over the first, and in various posi-

---

* An Experimental Inquiry into the Nature and Propagation of Heat. By J. Leslie. p. 534.

tions cuts off more or less of the opening. By means of this arrangement and of divisions engraved on the edge of each disk, it is easy to make the opening vary at will, and by a quantity which is perfectly definite.

As in my experiments on atmospheric air I had not noted the time, I began them again with this object in view ; and I found that the time of escape was always eleven seconds. This time did not vary with the density of the air ; and although this is the way it should be, it is none the less interesting to see the theory so well confirmed by experiment.

For hydrogen, I then diminished the opening until the time of escape was the same as that for atmospheric air. In spite of this equality of circumstances the changes in temperature were most different, as can be seen from the means of the results of four experiments :

Density of Hydrogen Expressed by Barometer.	Cold Produced in Flask No. 1.	Heat Produced in Flask No. 2.
$0^m.76$	$0°.92$	$0°.77$
$0^m.38$	$0°.54$	$0°.54$

The cold produced in the flask in which the hydrogen was, instead of being $0°.61$ as for atmospheric air, was $0°.92$; and the heat, instead of being $0°.58$, was found to be $0°.77$. The difference between $0°.92$ and $0°.77$ is much greater than that between $0°.61$ and $0°.58$; but as it is not probable that the changes in temperature for hydrogen follow a different law from those for air, I am inclined to believe that the difference between $0°.77$ and $0°.92$ is due solely to some condition of the experiment. We will see, in fact, that when the temperatures differ less in either direction from that of the surrounding medium, there is a better agreement.

The density of hydrogen being reduced one-half in the two flasks, I exhausted No. 2, and after the re-establishment of uniform temperature I opened communication with No. 1. I speak here of but one experiment; but it is, in fact. the mean of the results of four experiments which I consider. The heat absorbed was $0°.54$, and that liberated was also $0°.54$. This number is greater than one-half of $0°.92$, the figure obtained in the first experiment, and the difference is greater than that presented by the two corresponding numbers $0°.34$ and $0°.61$, of the experiments on atmospheric air, a fact which seems to me to confirm the idea that when the changes in temperature are very great, the errors are also the greatest. It seems to me, then, that when hydrogen changes its volume, owing to an increase or decrease of the weight which compresses it, the resulting changes in temperature obey the same law as do those experienced by air, but that the former are much greater.

I call to mind here that Mr. Leslie, whose work on heat contains some most beautiful experiments and many new ideas, was led into error in some way when he saw that hydrogen admitted into a receiver exhausted of air to nearly one-tenth produced the same effect as did air itself if admitted in its place. We have just seen that the changes of temperature which these two elastic fluids produce are most different, and that, consequently, the conclusion which he has drawn, that they contain in equal volumes the same quantity of heat, falls of itself.*

Having determined, as well as I could, the temperature changes which accompany those of density in hydrogen, I began the investigation of carbonic acid.

Having ascertained by preliminary trials the size of opening which gave a time of escape of eleven seconds, as in the case of hydrogen and air, I performed the experiments in the same way as for the other gases, and formed in the same manner the following table, which includes the means of the results of five experiments. It should be noticed that when carbonic acid rushed into the empty flask, it produced a loud, hissing sound. This is in general greater for gases of greater specific gravity.

Density of Carbonic Acid Gas Expressed by the Barometer.	Cold Produced in Flask No. 1.	Heat Produced in Flask No. 2.
$0^m.76$	$0°.56$	$0°.50$
$0^m.38$	$0°.30$	$0°.31$

The changes in temperature, either positive or negative, are nearly equal, as we see, and follow the law of densities ; but they are smaller than those for air, and, therefore, all the more so, smaller than those for hydrogen.

Likewise, oxygen has given, in a single experiment, it is true, but in one made with great care, the following results :

Density of Oxygen Expressed by the Barometer.	Cold Produced in Flask No. 1.	Heat Produced in Flask No. 2.
$0^m.76$	$0°.58$	$0°.56$
$0^m.38$	$0°.31$	$0°.32$

* An Experimental Inquiry, etc., p. 583.

Up to the present time I have been unable to extend my investigations further. If we compare, however, the results which we have obtained, we will be in a condition to deduce some new conclusions as a result of those already announced.

We see that, all circumstances being the same, the temperature changes produced by changes in volume are greater, the smaller the specific gravity of the gas. These changes are less for carbonic acid than for oxygen, less for oxygen than for atmospheric air, and much less for this last than for hydrogen, which is the lightest of all. Moreover, if we note that all gases are expanded to the same extent by heat, and that in our experiments, on occupying larger volumes (but the same for all), they absorbed quantities of heat which were inversely proportional to their specific gravities, we draw the important conclusion that the capacities for heat of gases, for equal volumes, increase as their specific gravities decrease.

My experiments have not yet given me the exact law of this connection. I think, however, that it may be determined; and I hope to make it the subject of a special investigation.

Of all known gases, hydrogen ought then to have the greatest capacity for heat, if I am not deceived by the results of my experiments. Since, further, oxygen and nitrogen differ only slightly in specific gravity, it would follow that they would have nearly the same capacity for heat. This is the reason why, in the memoir already quoted, on the analysis of air, we had found that these two gases stop the combustion of hydrogen at nearly the same point. This is also the reason why, as I have found recently, hydrogen stops it sooner than oxygen or nitrogen. It would be interesting to know exactly the influence of each gas in stopping the combustion of hydrogen; and I am hoping to make some new researches on this subject.

On collecting the various results which I have stated in this memoir, I think I can present as very probable the following consequences, which naturally spring from them:

1st. When a gas is made to occupy an empty space, the heat set free is not due to the traces of air which can be supposed to have been present.

2d. If we join two equal spaces, one empty, the other full of a gas, the thermometric changes which take place in each are the same.

3d. **For the same gas,** these temperature-changes **are** proportional to the changes in density which are experienced.

4th. These temperature-changes are not the same for all gases. They **are** greater the smaller **the** specific gravities.

5th. The capacities **of** the same gas **for heat** diminish with the density, **the** volume being the same.

6th. The capacities of **gases** for heat, **for equal volumes,** are greater the smaller the specific **gravities.**

I think I may repeat that I present these conclusions **only** with great reserve, knowing, myself, how **I need to** vary my experiments, and how **easy** it is to go astray in **the** interpretation of results; but although the new researches to **which** they lead me are immense, **I will not** allow myself to be stopped owing to their difficulty.

Biographical Sketch.

Louis Joseph Gay-Lussac was born **in St. Léonard, Limousin,** December 6, 1778, and died in Paris, **May 9, 1850.** He was educated **at the** École Polytechnique and **the** École des Ponts-et-Chaussées ; and became **Professor** of Chemistry at the École Polytechnique, **and Professor of** Physics in the Sorbonne.

His most important scientific researches were the **following :**

1. **The** discovery of simple **volume** relations in the **formation** of chemical compounds.

2. The laws of expansion **of gases with** change of temperature.

3. **The discovery of boron and** cyanogen, and his famous investigation **of** iodine.

4. The invention of **the** siphon-**barometer, and** numerous meteorological **investigations.**

On the Changes of Temperature Produced by the Rarefaction and Condensation of Air.

By JAMES PRESCOTT JOULE,

Phil. Mag., Series 3. XXVI., p. 369, **1845** ; *Scientific Papers*, Vol. I., p. 172.

CONTENTS

On the Changes of Temperature Produced by the Rarefaction and Condensation of Air.

By J. P. JOULE, Esq.*

In a paper† which was read before the Chemical Section of the British Association at Cork, I applied Dr. Faraday's fine discovery of magneto-electricity in order to establish definite relations between heat and the ordinary forms of mechanical power. In that paper it was demonstrated experimentally that the mechanical power exerted in turning a magneto-electrical machine is *converted into the heat* evolved by the passage of the currents of induction through its coils; and, on the other hand, that the motive power of the electro-magnetic engine is obtained at the expense of the heat due to the chemical reactions of the battery by which it is worked. I hope, at a future period, to be able to communicate some new and very delicate experiments, in order to ascertain the mechanical equivalent of heat with the accuracy which its importance to physical science demands. My present object is to relate an investigation in which I believe I have succeeded in successfully applying the principles before maintained to the changes of temperature arising from the alteration of the density of gaseous bodies—an inquiry of great interest in a practical as well as theoretical point of view, owing to its bearing upon the theory of the steam-engine.

Dr. Cullen and Dr. Darwin appear to have been the first who observed that the temperature of air is decreased by rarefaction and increased by condensation. Other philosophers have subsequently directed their attention to the subject. Dalton

---

* The experiments were made at Oak Field, Whalley Range, near Manchester.
† *Phil. Mag.*, Series 3, Vol. XXIII., pp. 263, 347, 435. [1843.]

was, however, the first who succeeded in measuring the change of temperature with some degree of accuracy. By the employment of an exceedingly ingenious contrivance, that illustrious philosopher ascertained that about 50° of heat are evolved when air is compressed to one-half of its original bulk, and that, on the other hand, 50° are absorbed by a corresponding rarefaction.*

There is every reason for believing that Dalton's results are very near the truth, especially as they have been exactly confirmed by the experiments of Dr. Ure with the thermometer of Breguet. But our knowledge of the specific heat of elastic fluids is of such an uncertain character that we should not be justified in attempting to deduce from them the absolute quantity of heat evolved or absorbed. I have succeeded in removing this difficulty by immersing my condensing-pump and receiver into a large quantity of water, so as to transfer the calorific effect to a body which is universally received as the standard of capacity.

My apparatus will be understood on inspecting Fig. 1. C represents the condensing-pump, consisting of a cylinder of gun-metal and of a piston fitted with a plug of oiled leather, which works easily, yet tightly, through a stroke of 8 inches. The cylinder is $10\frac{1}{2}$ inches long, $1\frac{3}{8}$ inches in interior diameter, and $\frac{1}{4}$ of an inch in thickness of metal. The pipe A, for the admission of air, is fitted to the lower part of the cylinder; at the bottom of this pipe there is a conical valve, constructed of horn, opening downward. A copper receiver, R, which is 12 inches long, $4\frac{1}{2}$ inches in exterior diameter, $\frac{1}{4}$ of an inch thick, and has a capacity of $136\frac{1}{2}$ cubic inches, may be screwed upon the pump at pleasure. This receiver is furnished with a conical valve of horn opening downward, and at the bottom with a piece of brass, B, along the centre of which there is a bore of $\frac{1}{4}$ of an inch diameter. There is a stopcock at S which I shall describe more particularly in the sequel.

Anticipating that the changes of temperature of the large quantity of water which was necessary, in order to surround the receiver and pump, would be very minute, I was at great pains in providing a thermometer of extreme sensibility and very

---

* *Memoirs of the Literary and Philosophical Society of Manchester*, Vol. V., Part 2, pp. 251–525.

great accuracy. A glass tube of narrow bore having been selected, a column of mercury, 1 inch long, was introduced, and gradually advanced in such a manner that the end of the

FIG. 1. *Scale,* $\frac{1}{12}$.

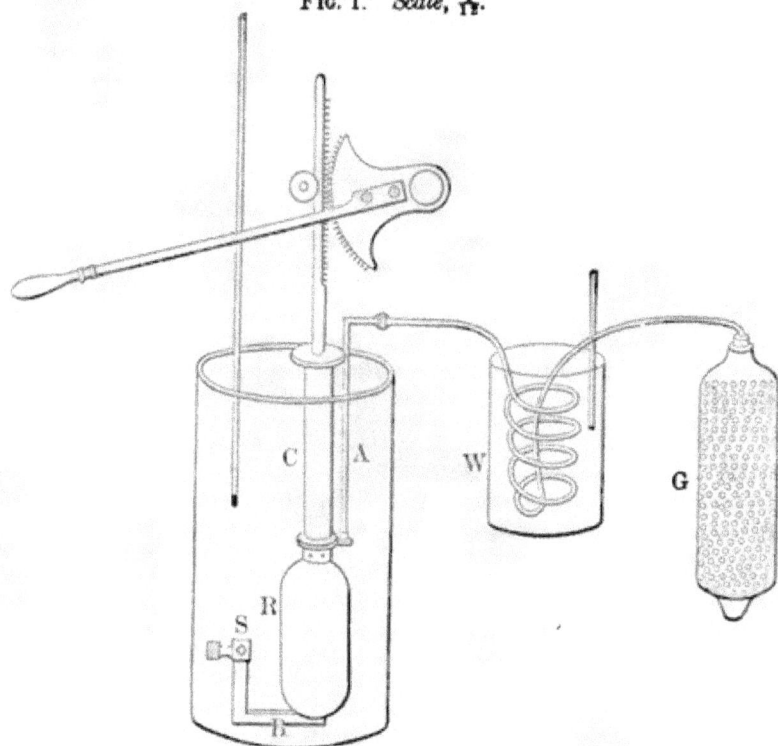

column in one position coincided with the beginning of the column in the next. In each position the length of the column was ascertained to the $\frac{1}{4000}$ part of an inch, by means of an instrument invented for the purpose by **Mr.** Dancer.* Afterwards the tube was covered with a film of beeswax, and each of the previously measured spaces was divided into twenty equal parts by means of a steel point carried by the dividing instrument; it was then etched by exposure to the vapor of fluoric

---

* Of the firm of Abraham & Dancer, Cross Street, Manchester. I have great pleasure in acknowledging here the skill displayed by this gentleman in the construction of the different parts of my apparatus. To it I must, in a great measure, attribute whatever success has attended the experiments detailed in this paper.

acid. The scale thus formed was entirely arbitrary ; and as it only extended between 30° and 90°, it was necessary to compare the thermometer with another, constructed in the same manner, but furnished with a scale including the boiling as well as the freezing point. When this was done it was found that ten divisions of the sensible thermometer (occupying about $\frac{1}{2}$ an inch) were nearly equal to the degree of Fahrenheit ; therefore, since by practice I can easily estimate with the naked eye $\frac{1}{20}$ of each of these divisions, I could with this instrument determine temperatures to the $\frac{1}{200}$ part of a degree. The scale being arbitrary, the indications of the thermometer had to be reduced in every instance, a circumstance which accounts for my having given the temperatures in the tables to three places of decimals.

It was important to employ, for the purpose of containing the water, a vessel as impermeable to heat as possible. With this view, two jars of tinned iron—one of them every way an inch smaller than the other—having been provided, the smaller jar was placed within the larger one, and the interstice between the two was closed hermetically. By this means a stratum of air of very nearly the same temperature as the water was kept in contact with the sides and bottom of the inner jar. The jars used in the other experiments which I shall bring forward were constructed in a similar manner. Among other precautions to insure accuracy, proper screens were placed between the vessels of water and the experimenter.

My first experiments were conducted in the following manner : The pump and copper receiver were immersed in 45 lb. 3 oz. of water, into which the very sensible thermometer above described was then placed, while two other thermometers were employed in order to ascertain the temperature of the room and that of the water contained by the vessel W. Having stirred the water thoroughly, its temperature was carefully read off. The pump was then worked at a moderate degree of speed until about 22 atmospheres of air, dried by being passed through the vessel G, full of small pieces of chloride of calcium, were compressed into the copper receiver. After this operation (which occupied from fifteen to twenty minutes) the water was stirred for five minutes so as to diffuse the heat equally through every part, and then its temperature was again read off. The increase of temperature thus observed was owing partly to the condensation of the air, and partly also to the friction of the pump and the motion of the water during the process of stirring. To estimate the value of the latter sources of heat, the air-pipe A was closed, and the pump was worked at the same velocity and for the same time as before, and the water was afterwards stirred precisely as in the first instance. The consequent increase of temperature indicated heat due to friction, etc.

The jar was now removed, and the receiver, having been immersed into a pneumatic trough, the quantity of air which had been compressed into it was measured in the usual manner and then corrected for the force of vapor, etc. The result added to 136.5 cubic inches, the quantity contained by the receiver at first, gave the whole quantity of compressed air.

The result given in Table I. is the difference between the effects of condensation and friction alone, corrected for the slight superiority of the cooling influence of the atmosphere in the experiments on friction. We must now, however, proceed to apply a further correction on account of the circumstance that the friction of the piston was considerably greater during the condensing experiments than during the experiments to ascertain the effect of friction. In the latter case the piston worked with a vacuum beneath it, while in the former the leather was pressed to the sides of the pump by a force of condensed air averaging 32 lb. per square inch. I endeavored to

TABLE I.

Source of Heat	Number of Strokes of Pump	Barometrical Pressure	Quantity of Air Compressed in Cubic Inches	Temp. of the Air Admitted	Mean Temp. of the Room	Difference	Temperature of Water Before Experiment	After Experiment	Heat Gained
				°	°	°	°	°	°
Condensation, etc..	300	30.06	3047	56.2	57.5	2.224−	54.930	55.622	0.692
Friction, etc........	300				57.5	1.885−	55.652	55.979	0.327
Condensation, etc..	300	30.07	2924	54.8	53.5	0.817+	53.970	54.664	0.694
Friction, etc........	300				54.5	0.358+	54.675	55.042	0.367
Condensation, etc..	300	30.24	2870	53.7	52.5	0.380+	52.562	53.197	0.635
Friction, etc........	300				52.6	0.760+	53.197	53.524	0.327
Condensation, etc..	300	30.07	2939	58.8	57.5	1.794−	55.559	56.053	0.494
Friction, etc........	300				57.75	1.536−	56.053	56.375	0.322
Condensation, etc..	300	29.34	2924	55.7	53.5	2.184+	55.409	55.959	0.550
Friction, etc........	300				53.75	2.316+	55.962	56.170	0.208
Condensation, etc..	300	30.80	3033	58.1	60.0	0.174+	59.876	60.472	0.596
Friction, etc........	300				60.4	0.196+	60.478	60.713	0.235
Condensation, mean	300	30.20	2936	56.2		0.078−			0.643
Friction, etc., mean	300					0.068+			0.297
Corrected Result...		30.20	2936						0.344

estimate the difference between the friction in the two cases by removing the valve of the receiver and working the pump with about 32 lb. per square inch pressure below it. These experiments, alternated with others in which a vacuum was beneath the piston, showed that the heat given out in the two cases was, as nearly as possible, in the ratio of six to five. When the correction indicated in this manner has been applied to 0°.297 (see Table) and the result subtracted from 0°.643, we obtain 0°.285 as the effect of compressing 2956 cubic inches of dry air at a pressure of 30.2 inches of mercury into the space of 136.5 cubic inches.

This heat was distributed through 45 lb. 3 oz. of water, 20½ lb. of brass and copper, and 6 lb. of tinned iron. It was therefore equivalent to 13°.628 per lb. avoirdupois of water.

The force necessary to affect the above condensation may be easily deduced from the law of Boyle and Mariotte, which has been proved by the French academicians to hold good as far as the twenty-fifth atmosphere of pressure. Let Fig. 2 represent

FIG. 2.

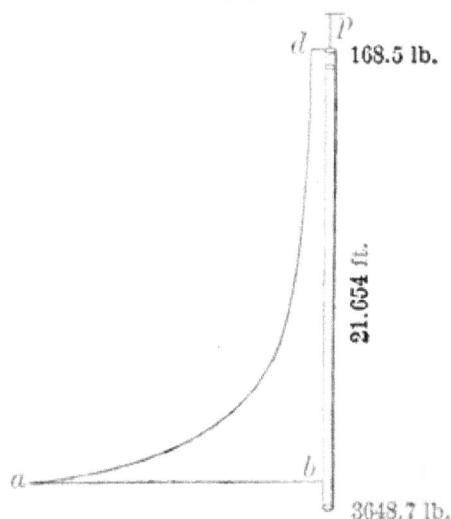

168.5 lb.

21.654 ft.

3648.7 lb.

a cylinder closed at one end, the length of which is 21.654 feet, and the sectional area 11.376 square inches. Then one foot of it will have exactly the same capacity as the copper receiver used in the experiments, and its whole capacity will be 2956 cubic inches. It is evident, therefore, that the force used in pumping (considered to be without friction) was exactly equal to that which would push the piston p to the distance of a foot from the bottom of the cylinder. Excluding exterior atmospheric pressure, the force upon the piston, when at the top of the cylinder, will be 168.5 lb., the weight of a column of mercury 30.2 inches long and of 11.376 square inches' section ; and at a foot from the bottom it will be 21.654 times as much, or 3648.7 lb. The hyperbolic area, $a\ b\ c\ d$, will therefore represent the force employed in the condensation, including the assistance of the atmospheric pressure. Applying the formula for hyperbolic spaces, we have

$$s = 3648.7 \times 2.302585 \times \log 21.654 = 11220.2.$$

The force expended in condensation was therefore equivalent to that which can raise 11220.2 lb. to the perpendicular height of one foot.

Comparing this with the quantity of heat evolved, we have $\frac{11220.2}{13.628} = \frac{823}{1}$ So that a mechanical force capable of raising 823 lb. to the height of one foot must be applied in the condensation of air in order to increase the temperature of a pound of water by one degree of Fahrenheit's scale.

The following table contains the results of experiments similar to the last, except in the extent to which the compression of air was carried.

TABLE II.

Source of Heat	Number of Strokes of Pump	Barometrical Pressure	Quantity of Air Compressed to Cubic Inches	Temp. of the Air Admitted	Mean Temp. of the Room	Difference	Temperature of Water		Heat Gained
							Before Experiment	After Experiment	
				°	°	°	°	°	°
Condensation, etc..	120	30.40	1410	54.0	54.2	0.010+	54.099	54.322	0.223
Friction, etc.......	120				54.6	0.224—	54.332	54.421	0.089
Condensation, etc..	120	30.50	1467	56.6	56.5	0.308+	56.693	56.923	0.230
Friction, etc.......	120				56.7	0.281+	56.926	57.036	0.110
Condensation, etc..	120	30.50	1440	62.6	63.6	1.763—	61.703	61.971	0.268
Friction, etc.......	120				64.0	1.960—	61.976	62.105	0.129
Condensation, etc..	120	30.57	1442	59.0	58.4	0.400+	58.680	58.921	0.241
Friction, etc.......	120				58.5	0.477+	58.921	59.033	0.112
Condensation, etc..	120	29.94	1405	55.2	57.0	1.566—	55.310	55.558	0.248
Friction, etc.......	120				57.2	1.573—	55.563	55.692	0.129
Condensation, mean	120	30.38	1433	57.5		0.522—			0.242
Friction, etc., mean	120					0.600—			0.114
Corrected Result.		30.38	1433						0.128

After applying the proper correction for the increase of friction during condensation, and reducing the result, as before, to the capacity of a pound of water, I find 5°.26 to be the mean quantity of heat evolved by compression of air in the above series of experiments.

The mechanical force expended in the condensation is represented in this instance by

$$s = 1779.3 \times 2.302585 \times \log 10.498 = 4183.46.$$

Hence the equivalent of a degree of heat per pound of water, as determined by the above series, is 795 lb. raised to the height of one foot.

The mechanical equivalents of heat derived from the foregoing experiments were so near 838* lb., the result of magnetical experiments in which "latent heat" could not be suspected to interfere in any way, as to convince me that the heat evolved was simply the manifestation in another form of the mechanical power expended in the act of condensation. I was still further confirmed in this view of the subject by the following experiments.

I provided another copper receiver (E, Fig. 3) which had a

FIG. 3.

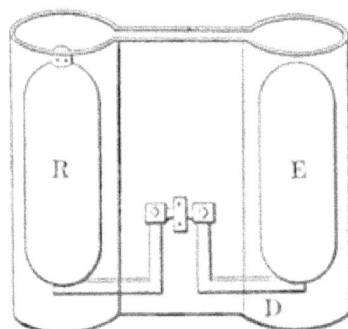

capacity of 134 cubic inches. Like the former receiver, to which it could be connected by a coupling nut, it had a piece D attached, in the centre of which there was a bore $\frac{1}{8}$ of an inch diameter, which could be closed perfectly by means of a proper stopcock.

* *Phil. Mag.*, Series 3, Vol. XXIII., p. 441.

I must here be permitted to make a short digression, in order to explain the construction of the stopcocks, as it may save those who shall in future attempt similar experiments the useless trouble of trying to make the ordinary stopcock perfectly air-tight under pressures. The one I have used is the invention of Mr. Ash, of this town, a gentleman well known for his great mechanical genius; and he has in the most obliging manner allowed me to give a full description of it. Fig. 4 is a full-

FIG. 4.

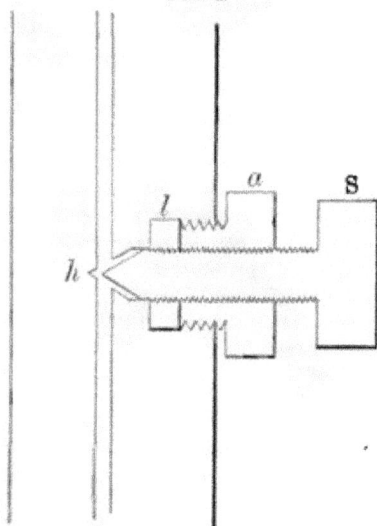

sized sectional view of the stopcock. *a* is a brass screw, by means of which a thick collar of leather, *l*, is very tightly compressed. The centre of *a* is perforated with a female screw, in which a steel screw, S, works, the threads of which press so tightly against the leather collar as effectually to prevent any escape of air in that direction. The end of the steel screw is smooth and conical, and the conical hole *h* is plugged with tin. When the stopcock is shut, the smooth end of the steel screw presses against the soft metal, so as to prevent the escape of the least particle of air; but when opened, as represented in the figure, it leaves a passage for the air around the conical point. I have tested this stopcock in the most severe manner, and have found it to answer perfectly.

Having filled the receiver R (Fig. 3) with about 22 atmospheres of dry air, and having exhausted the receiver E by means of an air-pump, I screwed them together, and then put them into a tin can containing 16½ lb. of water. The water was first thoroughly stirred, and its temperature taken by the same delicate thermometer which was made use of in the former experiments. The stopcocks were then opened by means of a proper key, and the air allowed to pass from the full into the empty receiver until equilibrium was established between the two. Lastly, the water was again stirred and its temperature carefully noted. The following table contains the results of a series of experiments conducted in this way, alternated with others to eliminate the effects of stirring, evaporation, etc. :

TABLE III.

Nature of Experiment	Barometrical Pressure	Quantity of Air Compressed in Receiver R in Cubic Inches	Mean Temperature of the Room	Difference	Temperature of Water		Gain or Loss of Heat
					Before Experiment	After Experiment	
			°	°	°	°	°
Expansion.........	30.20	2910	57.4	0.118+	57.520	57.517	0.003 loss
Alternation			57.0	0.906—	56.085	56.103	0.018 gain
Expansion.........	30.44	2920	57.0	0.885—	56.103	56.128	0.025 "
Alternation			62.0	0.783—	61.217	61.217	0
Expansion.........	30.44	2910	62.1	0.873—	61.222	61.232	0.010 "
Alternation			58.5	0.233+	58.732	58.735	0.003 "
Expansion.........	30.44	2915	58.6	0.132+	58.732	58.732	0
Alternation			61.3	0.787—	60.508	60.518	0.010 "
Expansion.........	30.46	3200	61.3	0.780—	60.518	60.523	0.005 "
Alternation			58.0	0.186+	58.184	58.187	0.003 "
Expansion.........	30.50	2880	58.3	0.110—	58.190	58.190	0
Mean of Experiments of Expansion	30.41	2956		0.400—			0.0062 gain
Mean of Alternations				0.411—			0.0064 gain
Corrected Result...	30.41	2956					0

The difference between the means of the expansions and alternations being exactly such as was found to be due to the increased effect of the temperature of the room in the latter case, we arrive at the conclusion that *no change of temperature occurs when air is allowed to expand in such a manner as not to develop mechanical power.*

In order to analyze the above experiments, I inverted the receivers, as shown in Fig. 5, and immersed them, as well as the connecting piece, into separate cans of water. One of the receivers had 2828 cubic inches of dry air condensed into it.

while the other was vacuous. After equilibrium was restored
by opening the cocks, I found that 2°.36 of cold per lb. of

Fig. 5.

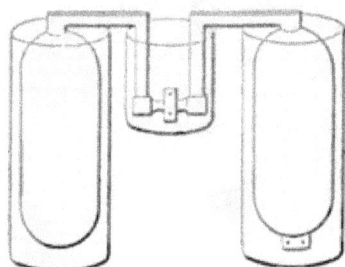

water had been produced in the receiver from which the air
had expanded, while 2°.38 of heat had been produced in the
other receiver, and 0°.31 of heat also in the can in which the
connecting piece was immersed, the sum of the whole amounting nearly to zero. The slight redundance of heat was owing
to the loss of cold during the passage of the air from the
charged receiver to the stopcocks, through a part of the pipe
which could not be immersed in water.

A series of experiments was now made in the following manner : The receiver was filled with dry, compressed air, and a
coiled leaden pipe, ¼ of an inch in internal diameter and 12
yards long, was screwed tightly upon the nozzle, as represented
in Fig. 6. The whole was then immersed into an oval can,
which was constructed as before described, and was also covered at top as perfectly as possible. Having ascertained the

Fig. 6.

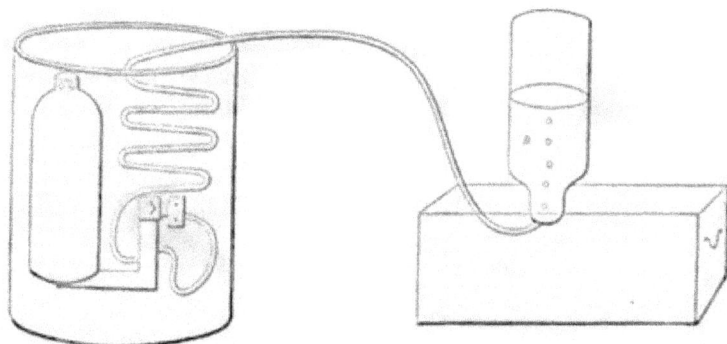

temperature of the water by means of the sensible thermometer before used, the stopcock was opened and the air made to pass from the receiver through a pneumatic trough into a jar, by which it was carefully measured. After the air in the receiver had been reduced to the atmospheric pressure, the water was again well stirred, and its temperature noted. An alternation was made after each of these experiments, in order to eliminate the effects of stirring, etc.

TABLE IV.

Nature of Experiment	Barometrical Pressure	Quantity of Air Compressed	Quantity of Air Let Out	Mean Temperature of the Room	Difference	Temperature of Water		Gain or Loss of Heat
						Before Experiment	After Experiment	
Expansion.....	30.04	2862	2726	55.7	0.405+	56.207	56.004	0.203 loss
Alternation				55.4	0.579+	56.004	55.954	0.050 loss
Expansion.....	30.10	2807	2670	54.6	0.022+	54.714	54.530	0.184 loss
Alternation				54.25	0.276+	54.536	54.516	0.020 loss
Expansion.....	30.10	2723	2587	53.6	0.760+	54.460	54.259	0.201 loss
Alternation				53.4	0.839+	54.259	54.219	0.040 loss
Expansion.....	30.10	2807	2670	49.05	0.307+	49.456	49.258	0.198 loss
Alternation				49.1	0.158+	49.258	49.258	0
Expansion.. ..	30.23	3039	2903	50.6	0.508+	50.176	50.008	0.168 loss
Alternation				51.1	1.063—	50.017	50.057	0.040 gain
Expansion. ...	30.20	2919	2782	49.0	0.355—	48.728	48.563	0.165 loss
Alternation				48.85	0.277—	49.573	48.573	0
Mean of Experiments of Expansion.	30.13	2859	2723		0.405+			0.1865 loss
Mean of Alternations .					0.085+			0.0117 loss
Corrected Result	30.13	2859	2723					0.1738 loss

The cold produced was diffused through 21.17 lb. of water, 14 lb. of copper, 8 lb. of lead, and 7 lb. of tinned iron. Hence we find that a quantity of cold was produced in the experiments sufficient to cause the temperature of a pound of water to decrease by 4°.085. At the same time a mechanical force was developed which could raise a column of the atmosphere of an inch square at the base to the altitude of 2723 inches; or, in other words, could raise 3352 lb. to the height of one foot. For each degree of heat lost there was therefore generated a force sufficient to raise 820 lb. to the height of one foot.

In the two following series the experiments were varied by

compressing and measuring out different volumes of air. [*Omitted.*]

These results are inexplicable if heat be a substance. If that were the case, the same quantity of heat would have been absorbed by the rarefaction which took place in the experiments of Table IV. as was evolved by the corresponding condensation in the experiments of Table I.; also a certain quantity of cold would have been produced in the experiments given in Table III. The results are, however, such as might have been deduced *a priori* from any theory in which heat is regarded as a state of motion among the constituent particles of bodies. It is easy to understand how the mechanical force expended in the condensation of air may be communicated to these particles so as to increase the rapidity of their motion, and thus may produce the phenomenon of increase of temperature. In the experiments of Table III. no cold was produced, because the momentum of these particles was not permanently converted into mechanical power, but had the motion of the air from one vessel to the other been opposed in such a manner as to develop power at the outside of the jar, which might have been accomplished by means of a cylinder and piston, then loss of heat would have occurred, just as in Tables IV., V., and VI., where the force was applied in lifting the atmosphere of the earth.

It is quite evident that the reason why the cold in the experiments of Table IV. was so much inferior in quantity to the heat evolved in those of Table I. is that all the force of the air over and above that employed in lifting the atmosphere was applied in overcoming the resistance of the stopcock, and was there converted back again into its equivalent of heat.

The discovery of Dulong[*] that. equal volumes of all elastic fluids, taken at the same temperature and under the same temperature, when suddenly compressed or dilated to the same fraction of their volume, disengage or absorb the same absolute quantity of heat accords perfectly with these principles.

The mechanical equivalents of heat determined by the various series of experiments given in this paper are 823, 795, 820, 814, and 760. The mean of the last three, which I take as least liable to error, is 798 lb., a result so near 838 lb., the equivalent which I deduced from my magnetical experiments, as to confirm in a remarkable manner the above explanation of the phenomena described in this paper, and to afford a new, and, to my mind, powerful argument in favor of the dynamical theory of heat which originated with Bacon, Newton, and Boyle, and has been at a later period so well supported by the experiments of Rumford, Davy, and Forbes. [*Two pages omitted.*]

Oak Field, near Manchester, *June*, 1844.

James Prescott Joule was born at Manchester, December 24, 1818, and died at Sale, his country-place, near Manchester, October 11, 1889. His father and grandfather were brewers; and he himself was in the business until 1854, when it was sold. He was sent as a boy to learn chemistry from Dalton, and became so interested in scientific investigation that his father furnished a laboratory for him at home. The researches described in the preceding paper were carried out in his own house at Oak Field, near Manchester. His later experiments were performed in the cellars of his house in Acton Square, Salford, and finally in a large yard attached to the brewery, New Bailey Street, Salford.

Among the most important of his researches may be mentioned :

The study of the heating effect of an electric current, and the discovery of the law $HJ = i^2 Rt$, which bears his name. (1840.)

The study of magnetization, lifting-power, saturation, changes of dimension, etc. (1841, 1846.)

The experimental verification of the identity of various forms of energy. (1843.)

Various researches on thermometry and measurement of temperature. He accurately determined the temperature of the maximum density of water.

A series of important experiments on gases. He was the first to substitute actual values in Laplace's corrected formula for the velocity of sound.

He was the first to define an *absolute* unit for electric current; also the first to use a *small* needle in a tangent galvanometer.

The memoirs which follow contain the most important investigations on the free expansion of gases carried out by Joule himself with the collaboration of Lord Kelvin, then William Thomson.

On the Thermal Effects of Fluids in Motion.

By WILLIAM THOMSON and J. P. JOULE.

PART I. *Phil. Trans*, 1853, Vol. CXLIII., p. 357; (omitted)
Abstract. *Proc. Roy. Soc.*, Vol. VI., 1853.

PART II. *Phil. Trans.*, 1854, Vol. CXLIV., p. 321;
Abstract. *Proc. Roy. Soc.*, Vol. VII., p. 127, 1854.

PART III. *Phil. Trans.*, 1860, Vol. CL., p. 325; (omitted)
Abstract. *Proc. Roy. Soc.*, Vol. X., p. 519, 1860.

PART IV. *Phil. Trans.*, 1862, Vol. CLII., p. 579;
Abstract. *Proc. Roy. Soc.*, Vol. XII., p. 202, 1862.

See also Joule's *Scientific Papers*, Vol. II., pp. 231-362; Thomson's *Mathematical and Physical Papers*, Vol. I., pp. 346-455.

CONTENTS

PART I.

PART II.

PART IV.

On the Thermal Effects of Elastic Fluids.

By PROFESSOR WILLIAM THOMSON, F.R.S., and J. P. JOULE,
Esq., F.R.S.

(Abstract of the preceding paper. [*Part I.*] *Proceedings Royal Society,*
June 16, 1853.)

THE authors had already proved, by experiments conducted
on a small scale, that when dry atmospheric air exposed to
pressure is made to percolate a plug of non-conducting porous
material, a depression of temperature takes place, increasing in
some proportion with the air in the receiver. The numerous
sources of error which were to be apprehended in experiments
of this kind, conducted on a small scale, induced the authors to
apply for the means of executing them on a larger scale ; and
the present paper contains the introductory part of their researches with apparatus furnished by the Royal Society, com-
prising a force-pump worked by a steam-engine, and capable of
propelling 260 cubic inches of air per second, and a series of
tubes by which the elastic fluid is conveyed through a bath of
water, by which its temperature is regulated, a flange at the
terminal permitting the attachment of any nozzle which is desired.

Preliminary experiments were made in order to illustrate the
thermal phenomena which result from the rush of air through
a single aperture. Two effects were anticipated—one of heat,
arising from the *vis viva* of air in rapid motion ; the other of
cold, arising from dilatation of the gas and the consequent
conversion of heat into mechanical effect. The latter was exhibited by placing the bulb of a very small thermometer close
to a small orifice through which dry atmospheric air, confined
under a pressure of eight atmospheres, was permitted to escape. In this case the thermometer was depressed 13° Cent.

below the temperature of the bath. The former effect was exhibited by causing the stream of air as it issued from the orifice to pass in a very narrow stream between the bulb of the thermometer and a piece of gutta-percha tube in which the latter was enclosed. In this experiment, with a pressure of eight atmospheres, an elevation of temperature equal to 23° Cent. was observed. The same phenomenon was even more strikingly exhibited by pinching the rushing stream with the finger and thumb, the heat resulting therefrom being insupportable.

The varied effects thus exhibited in the " rapids " neutralize one another at a short distance from the orifice, leaving, however, a small cooling effect, to ascertain the law of which and its amount for various gases the present researches have principally been instituted. A plug of cotton wool was employed for the purpose at once of preventing the escape of thermal effect in the rapids and of mechanical effect in the shape of sound. With this arrangement a depression of 0°.31 Cent. was observed, the temperature of the dry atmospheric air in the receiver being 14°.5 Cent., and its pressure 34.4 lbs. on the square inch, and the pressure of the atmosphere being 14.7 lbs. per square inch.

On the Thermal Effects of Fluids in Motion.

Part II.

By J. P. JOULE, F.R.S., and PROFESSOR W. THOMSON, M.A., F.R.S.* *(Phil. Trans.*, 1854, Vol. CXLIV., p. 321.)

(Plates I. and II.)

FIG. 1.

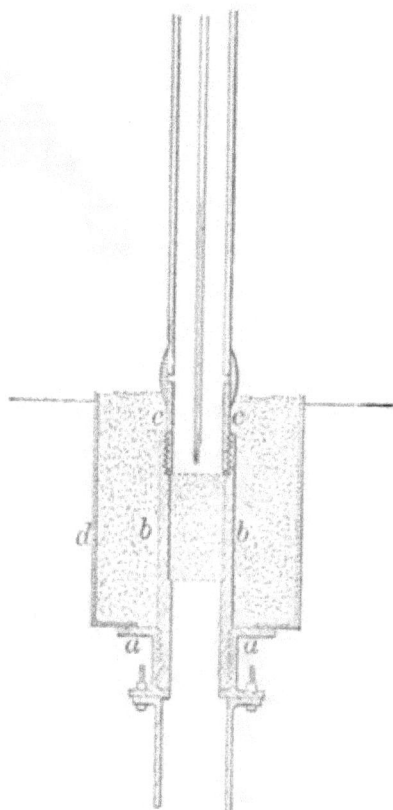

In the last experiment related in our former paper,† in which a low pressure of air was employed, a considerable variation of the cooling effect was observed, which it was necessary to account for in order to ascertain its influence on the results. We therefore continued the experiments at low pressure, trying the various arrangements which might be supposed to exercise influence over the phenomena. We had already interposed a plug of cotton-wool between the iron and copper pipes, which was found to have the very important effect of equalizing the pressure, besides stopping any solid or liquid particles driven from the pump, and which has therefore been retained in all the subsequent experiments. Another improvement was now effected by introducing a noz-

---

* The experiments were made at the Salford Brewery, New Bailey Street. † *Phil. Trans.*, 1853, Part III.

zle constructed of boxwood, instead of the brass one previously used. This nozzle is represented by Fig. 1, Plate 1, in which *a a* is a brass casting which bolts upon the terminal flange of the copper piping; *b b* is a turned piece of boxwood screwing into the above, having two ledges for the reception of perforated brass plates, the upper plate being secured in its place by the turned boxwood *c c*, which is screwed into the top of the first piece.

The space enclosed by the perforated plates is 2.72 inches long, and an inch and a half in diameter, and being filled with cotton, silk, or other material more or less compressed, presents as much resistance to the passage of the air as may be desired. A tin can, *d*, filled with cotton - wool, and screwing to the brass casting, serves to keep the water of the bath from coming in contact with the boxwood nozzle.

In the following experiments, made in order to ascertain the variations in the cooling effect above referred to, the nozzle was filled with 382 grs. of cotton-wool, which was sufficient to keep up a pressure of about 34 lbs. on the inch in the tubes when the pump was working at the ordinary rate. By opening the stopcock in the main pipe this pressure could be further reduced to about 22 lbs. by diminishing the quantity of air arriving at the nozzle. By shutting and opening the stopcock, we had therefore the means of producing a temporary variation of pressure, and of investigating its effect on the temperature of the air issuing from the nozzle. In the first experiments the stopcock was kept open for a length of time, until the temperature of the rushing air became pretty constant; it was then shut for a period of $3\frac{3}{4}$, $7\frac{1}{2}$, 15, 30, or 60 seconds, then reopened. The oscillations of temperature thus produced are laid down upon the Chart No. 1, in which the ordinates of the curves represent the temperatures according to the scale of thermometer C, each division corresponding to 0.0477 of a degree Centigrade. The divisions of the horizontal lines represent intervals of time equal to a quarter of a minute. The horizontal black lines show the temperature of the bath in each experiment. [*Chart is omitted.*]

The effect upon the pressure of the air produced by shutting the stopcock during various intervals of times, is given in the following table :

Stopcock Shut for	M.	S.	5 s.	15 s.	30 s.	1 m.	2 m.
Initial pressure			22.35	22.35	22.35	22.35	22.35
Pressure after	0	5	24.92	24.92	24.92	24.92	24.92
Pressure after	0	15	23.07	28.46	28.46	28.46	28.46
Pressure after	0	30	22.43	23.38	30.84	30.84	30.84
Pressure after	0	45	22.35	22.5	24.27	32.03	32.03
Pressure after	1	0	22.35	22.43	22.83	32.79	32.79
Pressure after	1	15		22.35	22.45	24.54	33.08
Pressure after	1	30		22.35	22.35	22.83	33.25
Pressure after	1	45			22.35	22.43	33.33
Pressure after	2	0				22.35	33.41
Pressure after	2	15				22.35	24.54
Pressure after	2	30					22.54
Pressure after	2	45					22.40
Pressure after	3	0					22.35

The last column gives also the effect occasioned by the permanent shutting or opening of the stopcock, 33.41 lbs. being nearly equal to the pressure when the stopcock has been closed for a long time.

In the next experiments, the opposite effect of opening the stopcock was tried, the results of which are laid down on Chart No. 2. [*Chart is omitted.*]

The effect upon the pressure of the air produced by opening the stopcock during the various intervals of time employed in the experiments is exhibited in the next table :

Stopcock Opened for	M.	S.	2½ s.	7½ s.	15 s.	30 s.	1 m.
Initial pressure			34.37	34.37	34.37	34.37	34.37
Pressure after	0	3¾	29.57	29.57	29.57	29.57	29.57
Pressure after	0	7½	.....	27.43	27.43	27.43	27.43
Pressure after	0	15	32.47	30.41	25.15	25.15	25.15
Pressure after	0	30	33.5	32.47	30.41	23.23	23.23
Pressure after	0	45	33.94	33.5	32.4	29.4	22.9
Pressure after	1	0	34.1	34.1	33.5	32.13	22.76
Pressure after	1	15	34.2	34.3	33.94	33.24	28.82
Pressure after	1	30	34.33	34.37	34.14	33.90	31.44
Pressure after	1	45	34.37	34.37	34.30	34.14	32.9
Pressure after	2	0			34.37	34.33	33.66
Pressure after	2	15				34.37	34.06
Pressure after	2	30					34.20
Pressure after	2	45					34.37

The remarkable fluctuations of temperature in the issuing stream accompanying such changes of pressure, and continuing to be very perceptible in the different cases for periods of from three to four minutes up to nearly half an hour after the pressure had become sensibly uniform, depend on a complication of circumstances, which appear to consist of (1) the change of cooling effect due to the instantaneous change of pressure; (2) a heating or cooling effect produced instantaneously by compression or expansion in all the air flowing towards and entering the plug, and conveyed through the plug to the issuing stream; and (3) heat or cold communicated by contact from the air on the high-pressure side to the metals and boxwood, and conducted through them to the issuing stream.

The first of these causes may be expected to influence the issuing stream instantaneously on any change in the stopcock; and after fluctuations from other sources have ceased it must leave a permanent effect in those cases in which the stopcock is permanently changed. But after a certain interval the reverse agency of the second cause, much more considerable in amount, will begin to affect the issuing stream, which will soon preponderate over the first, and (always on the supposition that this conviction is uninfluenced by conduction of any of the materials) will affect it with all the variations, undiminished in amount, which the air entering the plug experiences, but behind time by a constant interval equal to the time occupied by as much air as is equal in thermal capacity to the cotton of the plug in passing through the apparatus. This, in the experiments with the stopcock shut, would be very exactly a quarter of a minute; but it appears to have averaged more nearly one-third of a minute in the varying circumstances of the actual experiments, since our observations (as may be partially judged from the preceding charts) showed us, with very remarkable sharpness, in each case about twenty seconds after the shutting or opening of the stopcock, the commencement of the heating or cooling effect on the issuing stream, due to the sudden compression or rarefaction instantaneously produced in the air on the other side of the plug.

The entering air will, very soon after its pressure ceases to vary, be reduced to the temperature of the bath by the excellent conducting action of the spiral copper pipe through which it passes; and, consequently, twenty seconds or so later, the issuing stream can experience no further fluctuations in temperature except by the agency depending on the third cause.

That the third cause may produce very considerable effects is obvious, when we think how great the variations of temperature must be to which the surface of the solid materials in the neighborhood of the plug on the high-pressure side are subjected during the sudden changes of pressure, and that the heat consequently taken in or emitted by these bodies may influence the issuing stream perceptibly for a quarter or a half hour after the changes of pressure from which it originated have ceased, is quite intelligible on account of the slowness of conduction of heat through the wood and metals, when we take into account the actual dimensions of the parts of the apparatus round the plug. It is not easy, however, to explain all the fluctuations of temperature which have been observed after the pressure had become constant in the different cases. Those shown in the first set of diagrams are just such as might be expected from the alternate heating and cooling which the solids must have experienced at their surfaces on the high - pressure side, and which must be conducted through so as to affect the issuing stream after a considerable time; but the great elevations of temperature shown in the second set of diagrams, which correspond to cases when the pressure was temporarily or permanently *diminished,* are not, so far as we see, explained by the causes we have mentioned, and the circumstances of these cases require further examination.

When we had thus examined the causes of the fluctuations of temperature in the issuing air, the precautions to prevent their injurious effect upon the accuracy of the determinations of the cooling effect in the passage of air through the porous plug became evident. These were simply to render the action of the pump as uniform as possible, and to commence the record of observations only after one hour and a half or two hours had elapsed from the starting of the pump. The system then adopted was to observe the thermometers in the bath and stream of air, and the pressure - gauge every two minutes or minute and a half; the means of which observations are recorded in the columns of the tables. In some instances the air, previous to passing into the pump, was transmitted through a cylinder which had been filled with quick-lime. But since by previous use its power of absorbing water had been considerably deteriorated, a portion of the air was always transmitted through a Liebig tube containing asbestos moistened with sulphuric acid or chloride of zinc. The influence of a small quantity of moisture in the air is trifling, but will hereafter be examined. That of the carbonic acid contained by the atmosphere was, as will appear in the sequel, quite inappreciable. It will be proper to observe that the thermometers, by which the temperature of the bath and issuing air was ascertained, were repeatedly compared together to avoid any error which might arise from the alteration of their fixed points from time to time.

In each, excepting the first of the seven experiments recorded, the air passed through the quick-lime cylinder.

TABLE I.

Experiments with a plug consisting of 191 grains of cotton-wool.

1	2	3	4	5	6	7	8
No. of Observations from which the Results in Columns 4, 6, and 7 are Obtained	Cu. In. passed through Nozzle per Min.	Water in 100 Grains of Air, in Grains	Pressure in Lbs. on the Square Inch	Atmospheric Pressure	Temperature of the Bath	Temperature of the Issuing Air	Cooling Effect in Cent. Degrees
20	10822	0.51	21.326	14.400	20.295	20.201	0.094
20	10998	0.30	21.239	14.252	16.740	16.615	0.125
10	Not obsv'd	0.56	20.446	14.609	17.738	17.622	0.116
10	10769	0.66	20.910	14.772	16.039	15.924	0.115
10	10769	0.66	20.934	14.775	16.065	15.967	0.098
10	10769	0.66	20.995	14.779	16.084	15.984	0.100
10	10769	0.66	20.933	14.782	16.081	15.974	0.107
Mean	......	0.57	20.969	14.624	17.006	16.898	0.108

In the next experiments the nozzle was filled with 382 grains of cotton-wool. The intermediate stopcock, however, was partly opened, in order that by discharging a portion of air before its arrival at the nozzle the pressure might not be widely different from that employed in the last series. In all, excepting the last experiment recorded in the following table, the cylinder of lime was dispensed with.

TABLE II.

Experiments with a smaller quantity of air passed through a plug consisting of 382 grs. of cotton-wool.

1	2	3	4	5	6	7	8
No. of Observations from which the Results in Columns 4, 6, and 7 are Obtained	Cu. In. passed through Nozzle per Min.	Water in 100 Grains of Air, in Grains	Pressure in Lbs. on the Square Inch	Atmospheric Pressure	Temperature of the Bath	Temperature of the Issuing Air	Cooling Effect in Cent. Degrees
Mean	......	0.90	22.678	14.540	20.125°	19.979°	0.146°

TABLE III.

Experiments in which the entire quantity of air propelled by the pump was passed through a plug consisting of 382 grains of cotton-wool. The cylinder of lime was not employed.

1	2	3	4	5	6	7	8
No. of Observations from which the Results in Columns 4, 6, and 7 are Obtained	Cu. In. passed through Nozzle per Min.	Water in 100 Grains of Air, in Grains	Pressure in Lbs. on the Square Inch	Atmospheric Pressure	Temperature of the Bath	Temperature of the Issuing Air	Cooling Effect in Cent. Degrees
7	11766	0.56	36 625	14.583	19.869°	19.535°	0.334°
10	Not obsv'd	0.56	35.671	14.790	20.419	20.098	0.321
10	" "	0.36	35.772	14.504	16.096	15.730	0.366
10	" "	0.36	35.872	14.504	16.104	15.721	0.383
10	" "	0.36	36.026	14.504	16.232	15.869	0.363
Mean	......	0.44	35.993	14.577	17.744	17.390	0.354

In the next series of experiments the **air** was passed through a plug of silk, formed by rolling a silk handkerchief into a cylindrical shape, and then screwing it into the nozzle. The silk weighed 580 grains, and the small quantity of cotton-wool placed on the side next the thermometer, in order to equalize the stream of air more completely, weighed 15 grains. The stopcock was partly opened, as in the experiments of Table II., in **order** to reduce the pressure to that obtained by passing the full **quantity of air** propelled by the pump through a more porous **plug.** The cylinder of lime was employed.

TABLE IV.

Experiments in which a smaller quantity of air was passed through a plug consisting of 580 grains of silk.

1	2	3	4	5	6	7	8
No. of Observations from which the Results in Columns 4, 6, and 7 are Obtained	Cu. In. passed through Nozzle per Min.	Water in 100 Grains of Air, in Grains	Pressure in Lbs. on the Square Inch	Atmospheric Pressure	Temperature of the Bath	Temperature of the Issuing Air	Cooling Effect in Cent. Degrees
Mean	......	0.16	33.309	14.692	18.975°	18.610°	0.365°

TABLE V.

Experiments in which the entire quantity of air propelled by the pump was passed through the silk plug. The cylinder of lime was employed in all excepting the first two experiments.

1	2	3	4	5	6	7	8
No. of Observations from which the Results in Columns 4, 6, and 7 are Obtained	Cu. In. passed through Nozzle per Min.	Water in 100 Grains of Air, in Grains	Pressure in Lbs. on the Square Inch	Atmospheric Pressure	Temperature of the Bath	Temperature of the Issuing Air	Cooling Effect in Cent. Degrees
Mean	......	0.23	54.456	14.591	17.809°	17.102°	0.707°

In order to obtain a greater pressure, a plug was formed of silk "waste" compressed very tightly into the nozzle.

TABLE VI.

Experiments in which the air, after passing through the cylinder of lime, was forced through a plug consisting of 740 grains of silk.

1	2	3	4	5	6	7	8
No. of Observations from which the Results in Columns 4, 6, and 7 are Obtained	Cu. In. passed through Nozzle per Min.	Water in 100 Grains of Air, in Grains	Pressure in Lbs. on the Square Inch	Atmospheric Pressure	Temperature of the Bath	Temperature of the Issuing Air	Cooling Effect in Cent. Degrees
Mean	......	0.16	79.250	14.793	15.483°	14.373°	1.110°

In the foregoing experiments the pressure of the air on its exit from the plug was always exactly equal to the atmospheric pressure. To ascertain the effect of an alteration in the pressure of the exit air, we now enclosed a long siphon barometer within the glass tube (Fig. 14). The upper part of this tube was surmounted with a cap, furnished with a stopcock, by partially closing which the air at its exit could be brought to the required pressure. The influence of pressure in raising the mercury in the thermometer by compressing its bulb was ascertained by plunging the instrument into a bottle of water within the glass tube, and noting the amount of the sudden rise or fall of the quicksilver on a sudden augmentation or reduction of pressure. It was found that the pressure, equal to that of 17 inches of mercury, raised the indication by $0°.09$, which quantity was therefore subtracted after the usual reduction of the thermometric scale.

TABLE VII.

Experiments with the plug consisting of 740 grains of silk. Pressure of the exit air increased. Cylinder of lime used.

1	2	3	4	5	6	7	8
No. of Observations from which the Results in Columns 4, 6, and 7 are Obtained	Cu. In. passed through Nozzle per Min.	Water in 100 Grains of Air, in Grains	Pressure in Lbs. on the Square Inch	Atmospheric Pressure	Temperature of the Bath	Temperature of the Issuing Air	Cooling Effect in Cent. Degrees
Mean	Estimated at 5400	0.14	82.004	22.814	12.734°	11.701°	1.033°

With reference to the experiments in Table VII., it may be remarked that the cooling effect must be the excess of that which would have been obtained had the air been only resisted by the atmospheric pressure in escaping from the plug, above the cooling effect that would be found in an experiment with the temperature of the bath and the pressure of the entering air the same as the temperature and pressure of the exit air in the actual experiment, and the air issuing at atmospheric pressure. Hence, since two or three degrees of difference of temperature in the bath would not sensibly alter the cooling effect in any of the experiments on air, the cooling effect in an experiment in which the pressure of the exit air is increased must be sensibly equal to the difference of the cooling effects in two of the ordinary experiments, with the high pressures the same as those used for the entering and issuing air respectively, and the low pressure that of the atmosphere in each case ; a conclusion which is verified by the actual results, as the comparison given below shows.

The results recorded in the foregoing tables are laid down on Chart No. 3, in which the horizontal lines represent the excess of the pressure of the air in the receiver over that of the exit air as found by subtracting the fifth from the fourth columns of the tables, and the vertical lines represent the cooling effects in tenths of a degree Centigrade. It will be remarked that the line drawn through the points of observation is nearly straight, indicating that the cooling effect is, approximately at least, proportional to the excess of pressure, being about .018° per pound on the square inch of difference of pressure. Or we may arrive at the same conclusion by dividing the cooling effect (δ) by the difference of pressure $(P-P')$ in the different experiments. We thus find, from the means shown in the different tables :

$$\text{Table (I.)} \quad \frac{\delta}{P-P'} = .0170$$

$$
\begin{aligned}
\text{(II.)} &\quad .0179 \\
\text{(III.)} &\quad .0165 \\
\text{(IV.)} &\quad .0196 \\
\text{(V.)} &\quad .0177 \\
\text{(VI.)} &\quad .0172 \\
\text{(VII.)} &\quad \underline{.0174} \\
\text{Mean} &\quad .0176
\end{aligned}
$$

On the Cooling Effects Experienced by Carbonic Acid in Passing Through a Porous Plug.

The position of the apparatus gave us considerable practical facilities in experimenting with carbonic acid. A fermenting tun, 10 feet deep and 8 feet square, was filled with wort to a depth of 6 feet. After the fermentation had been carried on for about forty hours, the gas was found to be produced in sufficient quantity to supply the pump for the requisite time. The carbonic acid was conveyed by a gutta-percha pipe and passed through two glass vessels surrounded by ice in order to condense the greater portion of vapors. In the succeeding experiment the total quantity of liquid so condensed was 300 grains, which, having a specific gravity of .9965, was composed of 10 grains of alcohol and 290 grains of water. On analyzing a portion of the gas during the experiment by passing it through a tube containing chloride of zinc, it was found to contain 0.733 grain of water to 100 grains of carbonic acid.

In Table IX., as well as in the next series, the carbonic acid contained 0.35 per cent. of water.

In the experiment of Table X., as well as in those of the adjoining tables, the sudden diminution of pressure on connecting the pump with the receiver containing carbonic acid, is in perfect accordance with the discovery of Prof. Graham of the superior facility with which that gas may be transmitted through a porous body compared with an equal volume of atmospheric air.

TABLE VIII.

Carbonic acid forced through a plug of 382 grs. of cotton - wool. **Mean** barometric pressure, 29.45 inches, equivalent to 14.399 lbs. **Gauge** under atmospheric pressure, 151. The pump was placed **in connection** with the pipe immersed in carbonic acid at $10^h \cdot 55^m$.

Time of Observation	Vol. Percentage of Carbonic Acid	Pressure Gauge ; Mean Pressure in Lbs. on the Square Inch		Temperature of the Bath by Indications of Thermometer	Temperature of the Issuing Gas by Indications of Thermometer	Cooling Effect in Cent. Degrees
h. m.						
10 47	0	79.0		486.0		
49	0	79.0		486.0		
53	0	79.6				
57		85.2				
58		86.0				
59		85.0			188.6	
11 0	95.51	85.0			188.5	
2		86.4				
4		86.7				
6		86.6				
9	95.51	86.6				
13		84.0				
14		84.2	lbs.			
15	95.51 94.89	84.4	84.906 = 32.989	486.00 = $20^\circ.001$	188.36 = $18^\circ.611$	$1^\circ.390$
19		84.5				
22		84.1				
24		84.6				
25	93.03	84.2				
28		84.1				
32		83.2				
33		83.8				
35	86.82	84.0				
40		83.8				
41		83.9				
43		85.0				
45	79.37 80.61	86.0	84.245 = 32.286	485.94 = $19^\circ.998$	190.1 = $18^\circ.787$	$1^\circ.211$
49		84.6				
51		84.5				
53		83.9				
55	75.65	83.6				
12 0		83.6				
2		83.0				
5	70.68	82.7				
9	68.82	82.7	82.783 = 33.960	485.52 = $19^\circ.980$	191.07 = $18^\circ.884$	$1^\circ.096$
13		82.9				
15	66.96	82.7				
21		82.7				
23		82.8				
25	65.72	82.9				
28		82.9				
33		82.2				
35	63.23	82.3				
40		81.9				
44		81.9				
45	63.23 63.85	82.1	82.986 = 33.864	485.18 = $19^\circ.966$	191.82 = $18^\circ.959$	$1^\circ.007$
52		82.4				
55	62.0	83.9				
2		84.1				
5	63.23	84.9				
11		85.4				
15	65.72	82.1				

TABLE IX.

Carbonic acid forced through a plug consisting of 191 grs. of cotton-wool. Mean barometric pressure, 29.6 inches, equivalent to 14.472 lbs. Gauge under atmospheric pressure, 150.6. Pump placed in connection with the pipe immersed in carbonic acid at 10ʰ 38ᵐ.

1	2	3	4	5	6
Time of Observation	Vol. Percentage of Carbonic Acid	Pressure Gauge and Pressure in Lbs. on the Square Inch Equivalent Thereto	Indication of Thermometer, Temperature of the Bath	Indication of Thermometer, Temperature of the Issuing Gas	Cooling Effect in Cent. Degrees
h. m.					
10 40					
42					
44					
50	95.51				
53					
55		lbs.			
57	} 94.58	122.91 = 20.43	461.78 = 18°.962	187.49 = 18°.522	0°.44
59					
11 0	93.65				
1					
3					
5					
7					
9					
10	81.86				
11					
15	} 76.27	121.91 = 20.682	462.11 = 18°.976	188.35 = 18°.609	0°.367
17					
19					
20	79.68				
21					
25					

TABLE X.

Experiment in which carbonic acid was forced through a plug consisting of **580** grs. of silk. Mean barometric pressure, 29.56, equivalent to 14.452 lbs. Gauge under atmospheric pressure, 150.8. Pump placed in connection with the pipe immersed in carbonic acid at $12^h. 53^m$. Quantity of gas forced **through the plug,** about 7170 cubic **inches per minute.**

1	2	3	4	5	6
Time of Observation	Vol. Percentage of Carbonic Acid	Pressure Gauge and Pressure in Lbs. on the Square Inch Equivalent Thereto	Indication of Thermometer. Temperature of the Bath	Indication of Thermometer. Temperature of the Issuing Gas	Cooling Effect in Cent. Degrees
h. m.					
12 42	0	lbs.			
44	0	$52.2 = 55.454$	$464.34 = 19^\circ.072$	$185.53 = 18^\circ.323$	$0^\circ.749$
46	0				
49	0				
50	0				
52	0				
54					
57					
1 0	95.51				
5					
7					
9					
10	96.00 } 94.85	$55.92 = 51.7$	$464.47 = 19^\circ.077$	$165.0 = 16^\circ.256$	$2^\circ.821$
11					
13					
17					
20	93.03				
24					
25					
27					
30	85.92	$55.94 = 51.68$	$464.71 = 19^\circ.088$	$167.8 = 16^\circ.538$	$2^\circ.550$
35					
36					
38					

FREE EXPANSION OF GASES

TABLE XI.

Experiment in which carbonic acid was forced through a plug consisting of 740 grs. of silk. Mean barometric pressure, 30.065, equivalent to 14.723 lbs. on the inch. Gauge under atmospheric pressure 145.65. Pump placed in connection with the pipe immersed in carbonic acid at $11^h. 37^m$. Percentage of moisture in the carbonic acid, 0.15.

1	2	3	4	5	6
Time of Observation	Vol. Percentage of Carbonic Acid	Pressure Gauge and Pressure in Lbs. on the Square Inch Equivalent Thereto	Indication of Thermometer, Temperature of the Bath	Indication of Thermometer, Temperature of the Issuing Gas	Cooling Effect in Cent. Degrees
h. m.					
11 28		35.5			
30		35.1			
32		35.6			
34		35.2			
36		35.2			
37		36.0			
38		36.2			
39	95.51	36.6			
43		36.9			
45	95.51	37.0			
47		37.1			
50	95.51	37.0			
53		37.0			
55	95.51	37.0			
57		37.0	lbs.		
12 0	95.51	37.0	$37.0 = 75.324$	$319.17 = 12°.844$	$82.62 = 7°.974$	$4°.87$
2		37.0			
5	95.51	37.0			

(col 2: 95.51; col 3: 37.0)

In order **to ascertain** the cooling effect due to pure carbonic acid, we **may at present neglect** the effect due to the small quantity **of watery vapor contained** by **the** gas; and as the cooling effects observed **in the** various mixtures of atmospheric air and carbonic acid appear **nearly consistent** with the hypothesis that the specific heats of the two elastic fluids are for equal volumes equal to one another, and that each fluid experiences in the mixture the same thermo-dynamic effect as if the other were removed, **we** may for the present take the following **estimate of the** cooling effects **due to pure** carbonic acid, **at** the various temperatures and pressures employed, calculated by means of this hypothesis from the observations in which the percentage of carbonic **acid** was the greatest, and, in fact, so great that **a considerable error in** the correction for **the** common air would scarcely **affect the** result to any sensible extent.

D   49

	Temperature of the Bath	Excess of Pressure P-P'	Cooling Effect δ	Cooling Effect Divided by Excess of Pressure
From Table IX.	18.962	5.958	0.459	.0770
From Table VIII.	20.001	18.590	1.446	.0778
From Table X.	19.077	37.248	2.938	.0789
From Table XI.	12.844	60.601	5.049	.0833
	Mean 17.721		Mean of first three .0779	
			Mean of all .0793	

We shall see immediately that the temperature of the bath makes a very considerable alteration in the cooling effect, and we therefore select the first three results, obtained at nearly the same temperature, in order to indicate the effect of pressure. On referring to Chart No. 3, it will be remarked that these three results range themselves almost accurately in a straight line. Or, by looking to the numbers in the last column, we arrive at the same conclusion.

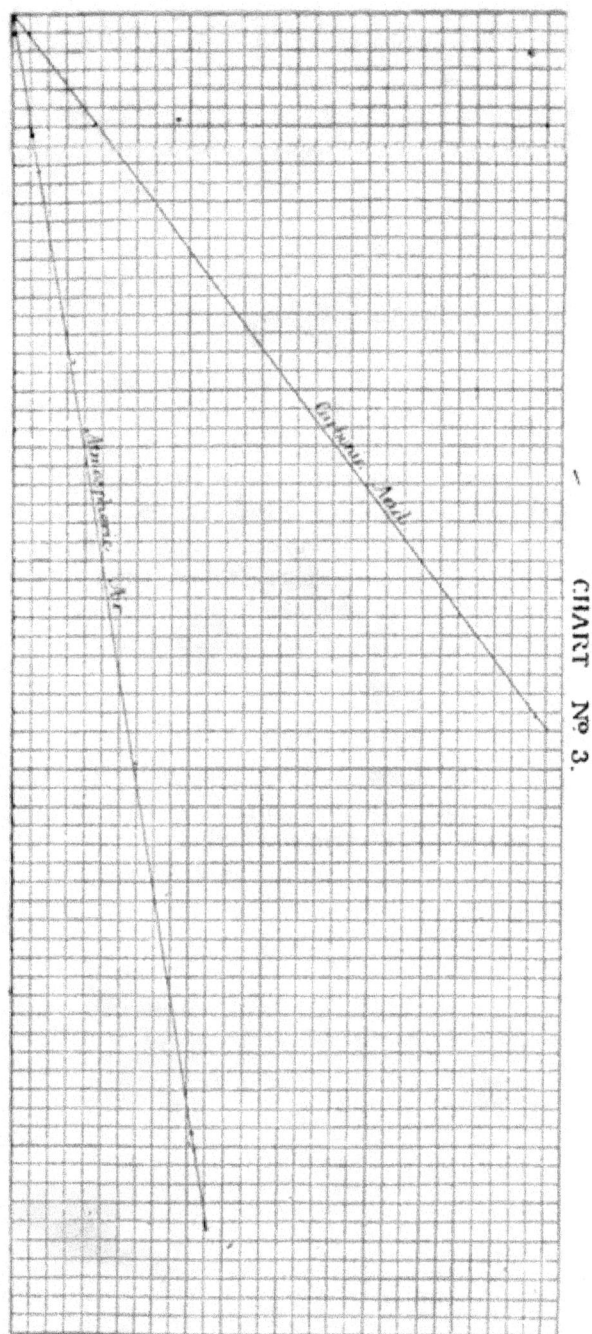

Carbonic Acid

Atmospheric Air

CHART № 3.

Cooling Effect * Experienced by Hydrogen in Passing Through a Porous Plug.

Not having been able as yet to arrange the large apparatus so as to avoid danger in using this gas in it, we have contented ourselves for the present with obtaining a determination by the help of the smaller force-pump employed in our preliminary experiments. The hydrogen, after passing through a tube filled with fragments of caustic potash, was forced, at a pressure of 68.4 lbs. on the inch, through a piece of leather in contact with the bulb of a small thermometer, the latter being protected from the water of the bath by a piece of india-rubber tube. At a temperature of about 10° Cent. a slight cooling effect was observed, which was found by repeated trials to be 0°.076. The pressure of the atmosphere being 14.7 lbs., it would appear that the cooling effect experienced by this gas is only one-thirteenth of that observed with atmospheric air. We state this result with some reserve, on account of the imperfection of such experiments on a small scale, but there can be no doubt that the effect of hydrogen is vastly inferior to that of atmospheric air.

Influence of Temperature on the Cooling Effect.

By passing steam through pipes plunged into the water of the bath we were able to maintain it at a high temperature without a considerable variation. The passage of hot air speedily raised the temperature of the stem of the thermometer, as well as of the glass tube in which it was enclosed; but nevertheless the precaution was taken of enclosing the whole in a tin vessel, by means of which water in constant circulation with the water of the bath was kept within one or two inches of the level of the mercury in the thermometer. The bath was completely covered with a wooden lid, and the water kept in constant and vigorous agitation by a proper stirrer.

[* *See Part IV.*]

TABLE XII.

Experiment in which, 1st, **air**; 2d, carbonic acid; 3d, air, **dried by quick-lime, was forced** through a plug consisting of 740 grs. of silk. Mean barometric pressure, 30.015, equivalent to 14.68 lbs. on the inch. Gauge under the atmospheric pressure 150. Percentage of moisture in the carbonic acid, 0.31. Pump placed in connection with the pipe immersed in carbonic acid at 11^h 24^m. Disconnected and attached to the quick-lime cylinder at 12^h 22^m.

Time of Observation		Vol. Percentage of Carbonic Acid		Pressure Gauge and Pressure in Lbs. on the Square Inch Equivalent Thereto	Indication of Thermometer. Temperature of the Bath	Indication of Thermometer. Temperature of the Issuing Gas	Cooling Effect in Cent. Degrees
H.	M.			lbs.			
11	5	0					
	7	0					
	9	0		31.62 = 91.508	646.15 = 91°.452	478.43 = 90°.008	1°.444
	11	0					
	13	0					
	15	0					
	17	0					
	19	0					
	21	0		31.95 = 90.576	646.68 = 91°.442	478.58 = 90°.043	1°.389
	22	0					
	23	0					
	24	0					
	25	0					
	26	0					
	30	95.51					
	32	95.51					
	33	95.51	95.51	32.23 = 89.799	646.59 = 91°.516	469.63 = 88°.044	3°.472
	36	95.51					
	38	95.51					
	40						
	43	93.03					
	46						
	48	90.60	91.81	32.1 = 90.162	647.03 = 91°.579	470.57 = 88°.255	3°.324
	50						
	53	80.82					
	55						
	58						
12	0	75.65	77.37	32.16 = 90.006	647.5 = 91°.647	472.29 = 88°.638	3°.009
	4						
	6	75.65					
	9						
	11	65.72					
	15	60.89	62.46	32.54 = 88.971	647.94 = 91°.711	474.64 = 89°.162	2°.549
	20	60.83					
	22						
	27	0					
	29	0					
	31	0					
	33	0					
	35	0					
	37	0					
	39	0					
	41	0					
	43	0		32.3 = 89.618	647.02 = 91°.578	480.97 = 90°.528	1°.050
	45	0					
	47	0					
	49	0					
	51	0					

Although hot air had been passed through the plug for half an hour before the readings in the preceding table were obtained, it is probable that the numbers 1.444 and 1.399, representing the cooling effect of atmospheric air, are not so accurate as the value 1°.050. Taking this latter figure for the effect of an excess of pressure of 89.618−14.68 = 74.938 lbs., we find a considerable decrease of cooling effect owing to elevation of temperature, for that pressure, at the low temperatures previously employed, is able to produce a cooling effect of 1°.309.

In order to obtain the effect of carbonic acid unmixed with atmospheric air, we shall, in accordance with the principle already adhered to, consider the thermal capacities of the gases to be equal for equal volumes. Then the cooling effect of the pure gas

$$= \frac{3.472 \times 100 - 1.052 \times 4.49}{95.51} = 3°.586.$$

Collecting these results we have :

Temperature of Bath	Excess of Pressure	Cooling Effect	Cooling Effect Reduced to 100 Lbs. Pressure	Theoretical Cooling Effect for 100 Lbs. Pressure
12.844	60.601	5 049	8.33	8.27
19.077	37.248	2.938	7.89	8.07
91.516	74.938	3.586	4.78	4.96

NOTE.—The numbers shown in the last column of the table are calculated by the general expression given in our former paper (*Phil. Trans.*, July, 1853) for the cooling effect, from an empirical formula for the pressure of carbonic acid, recently communicated by Mr. Rankine in a letter from which the following is extracted. [*Letter omitted.*]

The interpretation given above for the experimental results on mixtures of carbonic acid and air depends on the assumption (rendered probable as a very close approximation to the truth by Dalton's law) that in a mixture each gas retains all its physical properties unchanged by the presence of the other. This assumption, however, may be only approximately true, perhaps similar in accuracy to Boyle's and Gay-Lussac's laws of compression and expansion by heat ; and the theory of gases would be very much advanced by accurate comparative experiments on all the physical properties of mixtures and of their components separately. Towards this object we have experimented on the thermal effect of the mutual interpenetration of carbonic acid and air. In one experiment we found that when 7500 cubic inches of carbonic acid at the atmospheric pressure were mixed with 1000 cubic inches of common air, and a perfect mutual interpenetration had taken place, the temperature had fallen by about .2° Cent. We intend to try more exact experiments on this subject.

THEORETICAL DEDUCTIONS

SECTION 1.—*On the Relation between the Heat Evolved and the Work Spent in Compressing a Gas Kept at Constant Temperature.*

THIS relation is not a relation of simple mechanical equivalence, as was supposed by Mayer* in his *Bemerkungen ueber die Kräfte der Unbelebten Natur,* in which he founded on it an attempt to evaluate numerically the mechanical equivalent of the thermal unit. The heat evolved may be less than, equal to, or greater than the equivalent of the work spent, according as the work produces other effects in the fluid than heat, produces only heat, or is assisted by molecular forces in generating heat, and according to the quantity of heat, greater than, equal to, or less than that held by the fluid in its primitive condition, which it must hold to keep itself at the same temperature when compressed. The *a priori* assumption of equivalence, for the case of air, without some special reason from theory or experiment, is not less unwarrantable than for the case of any fluid whatever subjected to compression. Yet it may be demonstrated† that water below its temperature of maximum density (39°.1 Fahr.), instead of evolving any heat at all when compressed, actually absorbs heat, and at higher temperatures evolves heat in greater or less, but probably always very small, proportion to the equivalent of work spent; while air, as will be shown presently, evolves always, at least when kept at any temperature between 0° and 100° Cent., somewhat more heat

---

* *Annalen* of Wöhler und Liebig, May, 1842.
† "Dynamical Theory of Heat," § 63, equation (b), *Trans. Roy. Soc. Edinb.,* Vol. XVI., p. 290 ; or, *Phil. Mag.,* Vol. IV., Series 4, p. 425.

than the work spent in compressing it could alone create.
The first attempts to determine the relation in question, for the
case of air, established an approximate equivalence without
deciding how close it might be, or the direction of the discrepance, if any. Thus experiments " On the Changes of Temperature Produced by the Rarefaction and Condensation of Air," *
showed an approximate agreement between the heat evolved by
compressing air into a strong copper vessel under water, and
the heat generated by an equal expenditure of work in stirring
a liquid ; and again, conversely, an approximate compensation
of the cold of expansion when air in expanding spends all its
work in *stirring* its own mass by rushing through the narrow
passage of a slightly opened stop-cock. Again, theory,† without
any doubtful hypothesis, showed from Regnault's observations
on the pressure and latent heat of steam that, unless the density
of saturated steam differs very much from what it would be if
following the gaseous laws of expansion and compression, the
heat evolved by the compression of air must be sensibly less
than the equivalent of the work spent when the temperature is
as low as 0° Cent., and very considerably greater than the
equivalent when the temperature is above 40° or 50°. Mr.
Rankine is, so far as we know, the only other writer who inde-
pendently admitted the necessity of experiment on the subject,
and he was probably not aware of the experiments that had
been made in 1844, on the rarefaction and condensation of air,
when he remarked‡ that " the value of K is unknown ; and as
yet no experimental data exist by which it can be determined "
(K denoting in his expressions a quantity the vanishing of
which for any gas would involve the equivalence in question).
In further observing that K is probably small in comparison
with the reciprocal of the coefficient of expansion, Mr. Rankine
virtually adopted the equivalence as probably approximate ;
but in his article " On the Thermic Phenomena of Currents

* Communicated to the Royal Society, June 20, 1844, and published in
Philosophical Magazine, May, 1845.
† Appendix to " Account of Carnot's Theory," Roy. Soc. Edinb., April
30, 1849, *Transactions*, Vol. XVI., p. 568; confirmed in the " Dynamical
Theory," § 22, *Trans. Roy. Soc. Edinb.*, March 17, 1851; and *Phil. Mag.*, Vol.
IV., Series 4, p. 20.
‡ " Mechanical Action of Heat," Section II. (10), communicated to the
Roy. Soc. Edinb., Feb. 4, 1850 ; *Transactions*, Vol. XX., p. 166.

of Elastic Fluids,"[*] he took the first opportunity of testing it
closely, afforded by our preliminary experiments on the thermal
effects of air escaping through narrow passages.

We are now able to give much more precise answers to the
question regarding the heat of compression, and to others
which rise from it, than those preliminary experiments enabled
us to do. Thus if K denote the specific heat under constant
pressure, of air or any other gas issuing from the plug in the
experiments described above, the quantity of heat that would
have to be supplied, per pound of the fluid passing, to make
the issuing stream have the temperature of the bath would be
$K\delta$, or

$$K m \frac{(P - P')}{\Pi} \quad [\Pi \text{ is the atmospheric pressure}].$$

where m is equal to .26° for air and 1°.15 for carbonic acid, since
we found that the cooling effect was simply proportional to the
difference of pressure in each case, and was .0176° per pound
per square inch, or .26° per atmosphere, for air, and about $4\frac{1}{4}$
times as much for carbonic acid. This shows precisely how
much the heat of friction in the plug falls short of compensating the cold of expansion. But the heat of friction is the
thermal equivalent of all the work done actually in the narrow
passages by the air expanding as it flows through. Now this, in
the cases of air and carbonic acid, is really not as much as the
whole work of expansion, on account of the deviation from
Boyle's law to which these gases are subject; but it exceeds
the whole work of expansion in the case of hydrogen, which
presents a contrary deviation; since P'V', the work which a
pound of air must do to escape against the atmospheric pressure, is, for the two former gases, rather greater, and for hydrogen rather less, than PV, which is the work done on it in push-
ing it through the spiral up to the plug. In any case, w
denoting the whole work of expansion, $w - (P'V' - PV)$ will
be the work actually spent in friction within the plug; and

$$\frac{1}{J} \left\{ w - (P'V' - PV) \right\}$$

* "Mechanical Action of Heat," Subsection 4, communicated to the Roy.
Soc. Edinb., Jan. 4, 1853; Transactions, Vol. XX., p. 580.

will be the quantity of heat into which it is converted, a quantity which, in the cases of air and carbonic acid, falls short by

$$K m \frac{P - P'}{\Pi}$$

of compensating the cold of expansion. If, therefore, H denote the quantity of heat that would exactly compensate the cold of expansion, or which amounts to the same, the quantity of heat that would be evolved by compressing a pound of the gas from volume V' to the volume V, when kept at a constant temperature, we have

$$\frac{1}{J} \left\{ w - (P'V' - PV) \right\} = H - K m \frac{P - P'}{\Pi},$$

whence

$$H = \frac{w}{J} + \left\{ -\frac{1}{J} (P'V' - PV) + K m \frac{P - P'}{\Pi} \right\}.$$

Now from the results derived by Regnault from his experiments on the compressibility of air, of carbonic acid, and of hydrogen, at three or four degrees above the freezing-point, we find, approximately,

$$\frac{P'V' - PV}{PV} = f \frac{P - P'}{\Pi},$$

where
$$\quad f = \quad .00082 \text{ for air,}$$
$$\quad f = \quad .0064 \text{ for carbonic acid,}$$
and
$$\quad f = -.00043 \text{ for hydrogen.}$$

No doubt the deviations from Boyle's law will be somewhat different at the higher temperature (about 15° or 16° Cent.) of the bath in our experiments, probably a little smaller for air and carbonic acid, and possibly greater for hydrogen ; but the preceding formula may express them accurately enough for the rough estimate which we are now attempting.

We have, therefore, for air or carbonic acid :

$$H = \frac{w}{J} + \left(K m - \frac{PVf}{J} \right) \frac{P - P'}{\Pi} = \frac{w}{J} + \frac{PV}{J} \left(\frac{JKm}{PV} - f \right) \frac{P - P'}{\Pi}$$

The values of **JK** and PV for the three gases in the circumstances of the experiments are as follows :

For atmospheric air $JK = 1390 \times .238 = 331$;
For carbonic acid $JK = 1390 \times .217 = 301$;
For hydrogen $JK = 1390 \times 3.4046 = 4732$;

and for atmospheric air,

at 15° Cent. $PV = 26224 (1 + 15 \times .00366) = 27663$;

for carbonic acid,

at 10° Cent. $PV = 17154 (1 + 10 \times .00366) = 17782$;

for hydrogen,

at 10° Cent. $PV = 378960 (1 + 10 \times .00367) = 393000.$

Hence we have, for air and carbonic acid,

$$\text{II} = \frac{w}{J} + \frac{PV}{J} \cdot \lambda \frac{P - P'}{\text{II}},$$

where λ denotes .0024 for air, and .013 for carbonic acid, showing (since these values of λ are positive) that in the case of each of these gases more heat is evolved in compressing it than the equivalent of the work spent (a conclusion that would hold for hydrogen even if no cooling effect, or a heating effect less than a certain limit, were observed for it in our form of experiment). To find the proportion which this excess bears to the whole heat evolved, or to the thermal equivalent of the work spent in the compression, we may use the expression

$$w = PV \log \frac{P}{P'}$$

as approximately equal to the mechanical value of either of those energies ; and we thus find for the proportionate excess :

$$\frac{\text{II} - \frac{1}{J} w}{\frac{1}{J} w} = \lambda \frac{P - P'}{\text{II} \log \frac{P}{P'}} = .0024 \frac{P - P'}{\text{II} \log \frac{P}{P'}} \text{ for air,}$$

or

$$= .013 \frac{P - P'}{\text{II} \log \frac{P}{P'}} \text{ for carbonic acid.}$$

This equation shows in what proportion the heat evolved exceeds the equivalent of the work spent in any particular case of compression of either gas. Thus for a very small compression from $P' = \Pi$, the atmospheric pressure, we have

$$\log \frac{P}{P'} = \log\left(1 + \frac{P - \Pi}{\Pi}\right) = \frac{P - \Pi}{\Pi} \text{ approximately,}$$

and therefore

$$\frac{\Pi - \frac{1}{J}w}{\frac{1}{J}w} = .0024 \text{ for air}$$

or

$$= .013 \text{ for carbonic acid.}$$

Therefore, when slightly compressed from the ordinary atmospheric pressure, and kept at a temperature of about 60° Fahr., common air evolves more heat by $\frac{1}{417}$, and carbonic acid more by $\frac{1}{77}$, than the amount mechanically equivalent to the work of compression. For considerable compressions from the atmospheric pressure, the proportionate excesses of the heat evolved are greater than these values, in the ratio of the Napierian logarithm of the number of times the pressure is increased to this number diminished by 1. Thus, if either gas be compressed from the standard state to double density, the heat evolved exceeds the thermal equivalent of the work spent by $\frac{1}{210}$ in the case of air, and by $\frac{1}{33}$ in the case of carbonic acid.

As regards these two gases, it appears that the observed cooling effect was chiefly due to an actual preponderance of the mechanical equivalent of the heat required to compensate the cold of expansion over the work of expansion, but that rather more than one-fourth of it in the case of air, and about one-third of it in the case of carbonic acid, depended on a portion of the work of expansion going to do the extra work spent by the gas in issuing against the atmospheric pressure above that gained by it in being sent into the plug. On the other hand, in the case of hydrogen, in such an experiment as we have performed, there would be a heating effect if the work of expansion were precisely equal to the mechanical equivalent of the cold of expansion, since not only the whole work of expansion, but also the excess of the work done in forcing the gas in above that performed by it in escaping, is spent in friction in the plug. Since we have observed actually a cooling effect, it follows that the heat absorbed in expansion must exceed the equivalent of the work of expansion, enough to over-compensate the whole heat of friction mechanically equivalent, as this is, to the work of expansion, together with the extra **work** of sending the gas into the plug, above that which it **does in** escaping. In the actual experiment* **we found a** cooling **effect** of .076°, with **a** difference **of pressures, P—P'**, equal to 53.7 lbs. per square inch, or 3.7 atmospheres. Now the mechanical value of the **specific heat of** a pound of hydrogen is, according to the result stated above, 4732 foot-pounds, and hence the mechanical value of **the** heat that would compensate the observed cooling effect per pound of hydrogen **passing is 360** foot-pounds. But, according to Regnault's experiments **on** the compression of hydrogen, quoted above, we have

$$PV - P'V' = PV \times .00043 \, \frac{P - P'}{\Pi} \text{ approximately ;}$$

and as the temperature was about 10° in our **experiment, we** have, as stated above, $PV = 393000$.

Hence, for the **case of the** experiment in which **the** difference of pressure was 3.7 **atmospheres, or**

$$\frac{P - P'}{\Pi} = 3.7,$$

we have $$PV - P'V' = 625 ;$$

that is, 625 foot-pounds more **of work,** per pound of hydrogen, is spent in sending the hydrogen **into the** plug **at 4.7** atmospheres of pressures than would be gained in allowing it to escape at the same temperature against the atmospheric pressure. Hence,

---

*From the single experiment **we** have made on hydrogen we cannot conclude that at other pressures a cooling effect proportional to the difference of pressures would be observed, and therefore we confine the comparison of the three gases to the particular pressure used in the hydrogen experiment. It should be remarked, too, that we feel little confidence in the value assigned **to** the thermal effect for the case observed in the experiment on hydrogen, **and only** consider it established that it is a cooling effect, and **very small.**

the heat required to compensate the cold of expansion is generated by friction from (1) the actual work of expansion, together with (2) the extra work of 625 foot-pounds per pound of gas, and (3) the amount equivalent to 360 foot-pounds which would have to be communicated from without to do away with the residual cooling effect observed. Its mechanical equivalent, therefore, exceeds the work of expansion by 985 foot-pounds; which is $\frac{1}{630}$ of its own amount, since the work of expansion in the circumstances is approximately $393000 \times \log 4.7 = 608000$ foot-pounds. Conversely, the heat evolved by the compression of hydrogen at $10°$ Cent., from 1 to 4.7 atmospheres, exceeds by $\frac{1}{630}$ the work spent. The corresponding excess in the case of atmospheric air, according to the result obtained above, is $\frac{1}{174}$, and in the case of carbonic acid $\frac{1}{32}$.

It is important to observe how much less close is the compensation in carbonic acid than in either of the other gases, and it appears probable that the more a gas deviates from the gaseous laws, or the more it approaches the condition of a vapor at saturation, the wider will be the discrepancy. We hope, with a view to investigating further the physical properties of gases, to extend our method of experimenting to steam (which will probably present a large cooling effect), and perhaps to some other vapors.

In Mr. Joule's original experiment* to test the relation between heat evolved and work spent in the compression of air, without an independent determination of the mechanical equivalent of the thermal unit, air was allowed to expand through the aperture of an open stopcock from one copper vessel into another previously exhausted by an air-pump; and the whole external thermal effect on the metal of the vessels, and a mass of water below which they are kept, was examined. We may now estimate the actual amount of that external thermal effect, which observation only showed to be insensibly small. In the first place it is to be remarked that, however the equilibrium of pressure and temperature is established between the two air-vessels, provided only no appreciable amount of work is emitted in sound, the same quantity of heat must be absorbed from the

---

* The second experiment mentioned in the abstract published in the *Proceedings of the Royal Society,* June 20, 1844, and described in the *Philosophical Magazine,* May, 1845, p. 377. [*See p.* 17 *of this volume.*]

copper and water to reduce them to their primitive temperature ; and that this quantity, as was shown above, is equal to

$$\frac{PV}{J} \times 0.0024 \times \frac{P-P'}{\Pi} = \frac{27000 \times 0.0024}{1390} \times \frac{P-P'}{\Pi} = 0.046 \frac{P-P'}{\Pi}$$

In the actual experiments the exhausted vessel was equal in capacity to the charged vessel, and the latter contained .13 of a pound of air, under 21 atmospheres of pressure, at the commencement. Hence, $P' = \frac{1}{2}P$ and

$$\frac{P-P'}{\Pi} = 10.5 ;$$

and the quantity of heat required from without to compensate the total internal cooling effect must have been

$$.046 \times 10.5 \times .13 = .063.$$

This amount of heat taken from $16\frac{1}{2}$ lbs. of water, 28 lbs. of copper, and 7 lbs. of tinned iron, as in the actual experiment, would produce a lowering of temperature of only $.003°$ Cent. We need not therefore wonder that no sensible external thermal effect was the result of the experiment when the two copper vessels and the pipe connecting them were kept under water, stirred about through the whole space surrounding them, and that similar experiments, more recently made by M. Regnault, should have led only to the same negative conclusion.

If, on the other hand, the air were neither allowed to take in heat from nor to part with heat to the surrounding matter in any part of the apparatus, it would experience a resultant cooling effect (after arriving at a state of uniformity of temperature as well as pressure) to be calculated by dividing the preceding expression for the quantity of heat which would be required to compensate it by .17, the specific heat of air under constant pressure. The cooling effect on the air itself, therefore, amounts to

$$0°.27 \times \frac{P-P'}{\Pi},^{*}$$

* It is worthy of remark that this, the expression for the cooling effect experienced by a mass of atmospheric air expanding from a bulk in which

which is equal to 2°.8, for air expanding, as in Mr. Joule's experiment, from 21 atmospheres to half that pressure, and is 900 times as great as the thermometric effect when spread over the water and copper of the apparatus. Hence our present system, in which the thermometric effect on the air itself is directly observed, affords a test hundreds of times more sensitive than the method first adopted by Mr. Joule, and no doubt also than that recently practised by M. Regnault, in which the dimensions of the various parts of the apparatus (although not yet published) must have been on a corresponding scale, or in somewhat similar proportions, to those used formerly by Mr. Joule.

SECTION II.—*On the Density of Saturated Steam.*

The relation between the heat evolved and the work spent, approximately established by the air experiments communicated to the Royal Society in 1844, was subjected to an independent indirect test by an application of Carnot's theory, with values of "Carnot's function," which had been calculated from Regnault's data as to the pressure and latent heat of steam, and the assumption (in want of experimental data) that the density varies according to the gaseous laws. The verification thus obtained was very striking, showing an exact agreement with the relation of equivalence at a temperature a little above that of observation, and an agreement with the actual experimental results quite within the limits of the errors of observation; but a very wide discrepancy from equivalence for other temperatures. The following table is extracted from the appendix to the "Account of Carnot's Theory," in which the theoretical comparison was first made, to facilitate a comparison with what we now know to be the true circumstances of the case.

its pressure is P to a bulk in which, at the same (or very nearly the same) temperature, its pressure is P', and spending all its work of expansion in friction among its own particles, agrees very closely with the expression $26 \times \dfrac{P - P'}{H}$ for the cooling effect in the somewhat different circumstances of our experiments.

"Table of the Values of $\dfrac{\mu(1+Et)}{E}$" $= [W]$.

"Work requisite to produce a unit of heat by the compression of a gas $\dfrac{[\mu](1+Et)}{E} = [W]$	"Temperature of the gas t	"Work requisite to produce a unit of heat by the compression of a gas $\dfrac{[\mu](1+Et)}{E} = [W]$	"Temperature of the gas t
ft. lbs.	°	ft. lbs.	°
1357.1	0	1446.4	120
1368.7	10	1455.8	130
1379.0	20	1465.3	140
1388.0	30	1475.8	150
1395.7	40	1489.2	160
1401.8	50	1499.0	170
1406.7	60	1511.3	180
1412.0	70	1523.5	190
1417.6	80	1536.5	200
1424.0	90	1550.2	210
1430.6	100	1564.0	220
1438.2	110	1577.8	230 "

We know from the experiments described above in the present paper that the numbers in the first column, and, we may conclude with almost equal certainty, that the numbers in the third also, ought to be each very nearly the mechanical equivalent of the thermal unit. This having been ascertained to be 1390 (for the thermal unit Centigrade) by the experiments on the friction of fluids and solids, communicated to the Royal Society in 1849, and the work having been found above to fall short of the equivalent of the heat produced by about $\frac{1}{17}$, at the temperature of the air experiments at present communicated, and by somewhat less at such a higher temperature as 30°, we may infer that the agreement of the tabulated theoretical result with the fact is perfect at about 30° Cent. Or, neglecting the small discrepance by which the work truly required falls short of the equivalent of the heat produced, we may conclude that the true value of $\dfrac{\mu(1+Et)}{E}$ for all temperatures is about 1390; and hence that if $[W]$ denote the numbers shown for it in the preceding table, μ the true value of Carnot's function, and $[\mu]$ the value tabulated for any temperature in the "Account of Carnot's Theory," we must have to a very close degree of approximation,

$$\mu = [\mu] \times \frac{1390}{[W]}.$$

But if $[\sigma]$ denote the formerly assumed specific gravity of saturated steam, p its pressure, and λ its latent heat per pound of matter, and if ρ be the mass (in pounds) of water in a cubic foot, the expression from which the tabulated values of $[\mu]$ were calculated is

$$[\mu] = \frac{1-[\sigma]}{\rho[\sigma]} \frac{1}{\lambda} \frac{dp}{dt};$$

while the true expression of Carnot's function in terms of properties of steam is

$$\mu = \frac{1-\sigma}{\rho\sigma} \cdot \frac{1}{\lambda} \frac{dp}{dt}.$$

Hence,

$$\frac{\mu}{[\mu]} = \frac{[\sigma]}{\sigma} \frac{1-\sigma}{1-[\sigma]};$$

or, approximately, since σ and $[\sigma]$ are small fractions,

$$\frac{\mu}{[\mu]} = \frac{[\sigma]}{\sigma}.$$

We have, therefore,

$$\frac{\sigma}{[\sigma]} = \frac{[W]}{1390};$$

and we infer that the densities of saturated steam in reality bear the same proportions to the densities assumed, according to the gaseous laws, as the numbers shown for different temperatures in the preceding table bear to 1390. Thus we see that the assumed density must have been very nearly correct, about 30° Cent., but that the true density increases much more at the high temperatures and pressures than according to the gaseous laws, and consequently that steam appears to deviate from Boyle's law in the same direction as carbonic acid, but to a much greater amount, which, in fact, it must do unless its coefficient of expansion is very much less, instead of being, as it probably is, somewhat greater than for air. Also, we infer that the specific gravity of steam at 100° Cent., instead of being only $\frac{1}{1843.5}$, as was assumed, or about $\frac{1}{1700}$, as it is generally supposed to be, must be as great as $\frac{1}{1843}$. Without using the preceding table, we may determine the absolute density of saturated steam by means of a formula obtained as follows. Since we have seen the true value of W is nearly 1390, we must have, very approximately,

$$\mu = \frac{1390 \; E}{1 + Et},$$

and hence, according to the preceding expression for μ in terms of the properties of steam,

$$\rho\sigma = \frac{1-\sigma}{1390 \; E}(1 + Et)\frac{1}{\lambda}\frac{dp}{dt},$$

or, within the degree of approximation to which we are going (omitting as we do fractions such as $\frac{1}{400}$ of the quantity evaluated),

$$\rho\sigma = \frac{1 + Et}{1390 \; E\lambda}\frac{dp}{dt},$$

an equation by which $\rho\sigma$, the mass of a cubic foot of steam, in fractions of a pound, or τ, its specific gravity (the value of ρ being 63.887), may be calculated from observations such as those of Regnault on steam. Thus, using Mr. Rankine's empirical formula for the pressure which represents M. Regnault's own formula for the latent heat, and taking $E = \frac{1}{273}$, we have

$$\rho\sigma = \frac{273 + t}{1390} \; \frac{p\left(\dfrac{\beta}{(274.6 + t)^2} + \dfrac{2\gamma}{(274.6 + t)^3}\right) \times .4342945}{(606.5 + 0.305\, t) - (t + .00002\, t^2 + .0000003\, t^3)},$$

with the following equations for calculating p and the terms involving β and γ :

$$\log_{10} p = a - \frac{\beta}{t + 274.6} - \frac{\gamma}{(t + 274.6)^2},$$
$$a = 4.950433 + \log_{10} 2114 = 8.275538,$$
$$\log_{10} \beta = 3.1851091,$$
$$\log_{10} \gamma = 5.0827176.$$

The densities of saturated steam calculated for any temperatures, either by means of this formula, or by the expression given above, with the assistance of the table of values of [W], are the same as those which, in corresponding on the subject in 1848, we found would be required to reconcile Regnault's actual observations on steam with the results of air experiments which we then contemplated undertaking, should they turn out, as we now find they do, to confirm the relations which the air experiments of 1844 had approximately established. They should agree with results which Clausius* gave as a consequence of his extension of Carnot's principle to the dynamical theory of heat, and his assumption of Mayer's hypothesis.

Section III.—*Evaluation of Carnot's Function.*

The importance of this object, not only for calculating the efficiency of steam-engines and air-engines, but for advancing the theory of heat and thermo-electricity, was a principal reason for inducing us to undertake the present investigation. Our preliminary experiments, demonstrating that the cooling effect which we discovered in all of them was very slight for a considerable variety of temperatures (from about 0° to 77° Cent.), were sufficient to show, as we have seen in §§ I. and II., that $\frac{\mu (1 + Et)}{E}$ must be very nearly equal to the mechanical equivalent of the thermal unit; and therefore we have

$$\mu = \frac{J}{\frac{1}{E} + t} \text{ approximately,}$$

or, taking for E the standard coefficient of expansion of atmospheric air, .003665,

$$\mu = \frac{J}{272.85 + t}.$$

At the commencement of our first communication to the Royal

---

* Poggendorff's *Annalen*, April and May, 1850.

Society on the subject we proposed to deduce more precise values for this function by means of the equation

$$\frac{J}{\mu} = \frac{JK\delta - (P'V' - PV) + w}{\frac{dw}{dt}};$$

where

$$w = \int_V^{V'} p\,dv\,;$$

v, V, V' denote, with reference to air at the temperature of the bath, respectively, the volumes occupied by a pound under any pressure p, under a pressure P, equal to that with which the air enters the plug, and under a pressure P', with which the air escapes from the plug; and $JK\delta$ is the mechanical equivalent of the amount of heat per pound of air passing that would be required to compensate the observed cooling effect δ. The direct use of this equation for determining $\frac{J}{\mu}$ requires, besides our own results, information as to compressibility or expansion, which is as yet but very insufficiently afforded by direct experiments, and is consequently very unsatisfactory, so much so that we shall only give an outline, without details, of two plans we have followed, and mention the results. First, it may be remarked that, approximately,

$$w = (1 + \mathrm{E}t)\,\mathrm{H}\log\frac{\mathrm{P}}{\mathrm{P}'}\cdot \text{ and } \frac{dw}{dt} = \mathrm{E}\mathrm{H}\log\frac{\mathrm{P}}{\mathrm{P}'},$$

H being the "height of the homogeneous atmosphere," or the product of the pressure into the volume of a pound of air, at 0° Cent.; of which the value is 26224 feet. Hence if E denote a certain mean coefficient of expansion suitable to the circumstances of each individual experiment, it is easily seen that $\frac{w}{\frac{dw}{dt}}$ may be put under the form $\frac{1}{\mathrm{E}} + t$;

and thus we have

$$\frac{J}{\mu} = \frac{1}{E} + t + \frac{JK\delta - (P'V' - PV)}{EH \log \frac{P}{P'}},$$

since the numerator of the fraction constituting the last term is so small that the approximate value may be used for the denominator. The first term of the second member may easily be determined analytically in general terms; but as it has reference to the rate of expansion at the particular temperature of the experiment, and not to the mean expansion from 0° to 100°, which alone has been investigated by Regnault and others who have made sufficiently accurate experiments, we have not data for determining its values for the particular cases of the experiments. We may, however, failing more precise data, consider the expansion of air as uniform from 0° to 100° for any pressure within the limits of the experiments (four or five atmospheres), because it is so for air at the atmospheric density by the hypothesis of the air-thermometer; and Regnault's comparisons of air-thermometers in different conditions show for all, whether on the constant-volume or constant-pressure principle, with density or pressure from one - half to double the standard density or pressure, a very close agreement with the standard air-thermometer. On this assumption, then, when we take into account Regnault's observations regarding the effect of the variations of density on the coefficient of increase of pressure, we find that a suitable mean coefficient E for the circumstances of the preceding formula for $\frac{J}{\mu}$ is expressed to a sufficient degree of approximation by the equation

$$E = .0036534 + \frac{.0000441}{3.81} \frac{P - P'}{H \log \frac{P}{P'}}.$$

Also by using Regnault's experimental results on compressibility of air as if they had been made, not at 4°.75, but at 16° Cent., we have estimated $P'V' - PV$ for the numerator of the last term of the preceding expression. We have thus obtained estimates for the value of $\frac{J}{\mu}$ from eight of our experiments (not corresponding exactly to the arrangement in seven series given above), which, with the various items of the correction in the case of each experiment, are shown in the following table:

No. of Exp.	Pressure of Air Forced into the Plug	Barometric Pressure	Excess	Cooling Effect	Correction by Cooling Effect	Correction by Reciprocal Coefficient of Expansion	Correction by Compressibility (Subtracted)	Value of J divided by Carnot's Function for 16° Cent.
P	P'	P — P'	δ	$\dfrac{JK\delta}{EH \log \frac{P}{P'}}$	$\dfrac{1}{E} - \dfrac{1}{E}$	$\dfrac{P'V'-PV}{EH \log \frac{P}{P'}}$	$\dfrac{J}{\mu_{16}}$
I.	20.943	14.777	6.166	0.105	1.031	0.174	0.290	289.4
II.	21.282	14.326	6.956	0.109	0.942	0.168	0.291	289.3
III.	35.822	14.504	21.318	0.375	1.421	0.519	0.412	289.97
IV.	33.310	14.692	18.618	0.364	1.523	0.470	0.372	290.065
V.	55.441	14.610	40.831	0.740	1.892	0.923	0.480	289.705
VI.	53.471	14.571	38.900	0.676	1.814	0.883	0.475	289.59
VII.	79.464	14.955	64.509	1.116	2.272	1.379	0.592	289.69
VIII.	79.967	14.785	65.182	1.142	2.300	1.376	0.586	289.73
							Mean...	289.68

In consequence of the approximate equality of $\dfrac{J}{\mu}$ to $\dfrac{1}{E} + t$, its value must be, within a very minute fraction, less by 16 at 0° than at 16°; and from the mean result of the table we therefore deduce 273.68 as the value of $\dfrac{J}{\mu}$ at the freezing-point. The correction thus obtained on the approximate estimate $\dfrac{1}{E} + t = 272.85 + t$, for $\dfrac{J}{\mu}$, at temperatures not much above the freezing-point, is an augmentation of .83.

For calculating the unknown terms in the expression for $\dfrac{J}{\mu}$ we have also used Mr. Rankine's formula for the pressure of air, which is as follows:

$$pv = H \frac{C + t}{C} \left\{ 1 - \frac{aC}{(C+t)^2} \left(\frac{1}{\rho v} \right)^{\frac{2}{3}} + \frac{hC}{C + t} \left(\frac{1}{\rho v} \right)^{\frac{4}{3}} \right\},$$

where

$$C = 274.6, \quad \log_{10} a = .3176168, \quad \log_{10} h = 3.8181546,$$
$$H = \frac{26224}{1 - a + h};$$

and, v being the volume of a pound of air when at the temperature t and under the pressure p, ρ denotes the mass in pounds of a cubic foot at the standard atmospheric pressure of 29.9218 inches of mercury. The value of p according to this equation, when substituted in the general expression for $\dfrac{J}{\mu}$ gives:

$$\frac{J}{\mu} = C + t +$$

$$\frac{\dfrac{J K C}{H}\tilde{c} + 3\lambda \dfrac{C^{\frac{3}{2}}}{(C+t)^{\frac{1}{2}}}\left\{\left(\dfrac{P}{H}\right)^{\frac{1}{2}} - \left(\dfrac{P}{H}\right)^{\frac{1}{2}}\right\} - \dfrac{13}{3}a\left(\dfrac{C}{C+t}\right)^{\frac{3}{2}}\left\{\left(\dfrac{P}{H}\right)^{\frac{3}{2}} - \left(\dfrac{P}{H}\right)^{\frac{3}{2}}\right\}}{\log \dfrac{P}{P}}$$

From this we find, with the data of the eight experiments just quoted, the following values of $\dfrac{J}{\mu}$ at the temperature of $16°$ Cent.,

289.044, 289.008, 288.849, 289.112, 288.787, 288.722, 288.505, 288.559, the mean of which is 288.82,

giving a correction of only .03 to be subtracted from the previous approximate estimate $\dfrac{1}{E} + t$.

It should be observed that Carnot's function varies only with the temperature; and, therefore, if such an expression as the preceding, derived from Mr. Rankine's formula, be correct, the cooling effect, δ, must vary with the pressure and temperature in such a way as to reduce the complex fraction, constituting the second term, to either a constant or a function of t. Now at the temperature of our experiments, δ is very approximately proportional to simply $P - P'$, and therefore all the terms involving the pressure in the numerator ought to be either linear or logarithmic; and the linear terms should balance one another so as to leave only terms which, when divided by $\log \dfrac{P}{P'}$, become independent of the pressures. This condition is not fulfilled by the actual expression, but the calculated results agree with one another as closely as could be expected from a formula obtained with such insufficient experimental data as Mr. Rankine had for investigating the empirical forms which his theory left undetermined. We shall see in Section V. below that simpler forms represent Regnault's data within their limits of error of observation, and at the same time may be reduced to consistency in the present application.

As yet we have no data regarding the cooling effect of sufficient accuracy for attempting an independent evaluation of Carnot's function for other temperatures. In the following section, however, we propose a new system of thermometry, the adoption of which will quite alter the form in which such a problem as that of evaluating Carnot's function for any temperature presents itself.

SECTION IV.—*On an Absolute Thermometric Scale Founded on the Mechanical Action of Heat.*

In a communication to the Cambridge Philosophical Society * six years ago it was pointed out that any system of thermometry founded either on equal additions of heat, or equal expansions, or equal augmentations of pressure, must depend on the particular thermometric substance chosen, since the specific heats, the expansions, and the elasticities of substances vary, and, so far as we know, not proportionally with absolute rigour for any two substances. Even the air-thermometer does not afford a *perfect standard* unless the precise constitution and physical state of the gas used (the density, for a pressure-thermometer, or the pressure, for an expansion-thermometer) be prescribed ; but the very close agreement which Regnault found between different air- and gas-thermometers removes, for all practical purposes, the inconvenient narrowness of the restriction to atmospheric air kept permanently at its standard density, imposed on the thermometric substance in laying down a rigorous definition of temperature. It appears, then, that the standard of practical thermometry consists essentially in the reference to a certain numerically expressible quality of a particular substance. In the communication alluded to, the question, " Is there any principle on which an absolute thermometric scale can be founded ?" was answered by showing that Carnot's func-

* " On an Absolute Thermometric Scale Founded on Carnot's Theory of the Motive Power of Heat, and Calculated from Regnault's Observations on Steam," by Professor W. Thomson ; *Proceedings Camb. Phil. Soc.*, June 5, 1848, or *Philosophical Magazine*, October, 1848.

tion (derivable from the properties of any substance whatever, but the same for all bodies at the same temperature), or any arbitrary function of Carnot's function, may be defined as temperature, and is therefore the foundation of an absolute system of thermometry. We may now adopt this suggestion with great advantage, since we have found that Carnot's function varies very nearly in the inverse ratio of what has been called "temperature from the zero of the air-thermometer"—that is, Centigrade temperature by the air-thermometer increased by the reciprocal of the coefficient of expansion; and we may define temperature simply as the reciprocal of Carnot's function. When we take into account what has been proved regarding the mechanical action of heat,* and consider what is meant by Carnot's function, we see that the following explicit definition may be substituted :

If any substance whatever, subjected to a perfectly reversible cycle of operations, takes in heat only in a locality kept at a uniform temperature, and emits heat only in another locality kept at a uniform temperature, the temperatures of these localities are proportional to the quantities of heat taken in or emitted at them in a complete cycle of the operations.

To fix on a unit or degree for the numerical measurement of temperature, we may either call some definite temperature, such as that of melting ice, unity, or any number we please ; or we may choose two definite temperatures, such as that of melting ice and that of saturated vapour of water under the pressure 29.9218 inches of mercury in the latitude 45°, and call the difference of these temperatures any number we please—100, for instance. The latter assumption is the only one that can be made conveniently in the present state of science, on account of the necessity of retaining a connection with practical thermometry as hitherto practised ; but the former is far preferable in the abstract, and must be adopted ultimately. In the meantime it becomes a question, What is the temperature of melting ice, if the difference between it and the standard boiling-point be called 100° ? When the question is answered within a tenth of a degree or so, it may be convenient to alter the

* *Dynamical Theory of Heat,* §§ 42, 43.

foundation on which the degree is defined by assuming the temperature of melting ice to agree with that which has been found in terms of the old degree; and then to make it an object of further experimental research, to determine by what minute fraction the range from freezing to the present standard boiling point exceeds or falls short of 100. The experimental data at present available do not enable us to assign the temperature of melting ice, according to the new scale, to perfect certainty within less than two or three tenths of a degree; but we shall see that its value is probably 273.7, agreeing with the value $\frac{J}{\mu}$ at $0°$ found by the first method in Section III.

From the very close approximation to equality between $\frac{J}{\mu}$ and $\frac{1}{,E} + t$, which our experiments have established, we may be sure that the temperature from the freezing-point by the new system must agree to a very minute fraction of a degree with Centigrade temperature between the two prescribed points of agreement, $0°$ and $100°$, and we may consider it as highly probable that there will also be a very close agreement through a wide range on each side of these limits. It becomes, of course, an object of the greatest importance, when the new system is adopted, to compare it with the old standard; and this is, in fact, what is substituted for the problem, the evaluation of Carnot's function, now that it is proposed to call the reciprocal of Carnot's function, temperature. In the next section we shall see by what kind of an examination of the physical properties of air this is to be done, and investigate an empirical formula expressing them consistently with all the experimental data as yet to be had, so far as we know.

The following table, showing the indications of the constant-volume and constant-pressure air-thermometer in comparison for every twenty degrees of the new scale, from the freezing-point to $300°$ above it, has been calculated from the formulæ (9), (10), and (39) of Section V. below.

Temperature by Absolute Scale in Cent. Degrees from the Freezing point $t - 273.7$	Temperature Centigrade by Constant-volume Thermometer with Air of Specific Gravity $\dfrac{\Phi}{v}$ $\theta = 100 \dfrac{p_t - p_{273.7}}{p_{373.7} - p_{273.7}}$	Temperature Centigrade by Constant-pressure Air-thermometer $\mathfrak{s} = 100 \dfrac{v_t - v_{273.7}}{v_{373.7} - v_{273.7}}$
0°	0°	0°
20	$20 + .0298 \times \dfrac{\Phi}{c}$	$20 + .0404 \times \dfrac{p}{\Pi}$
40	40 + .0403 "	40 + .0477 "
60	60 + .0366 "	60 + .0467 "
80	80 + .0223 "	80 + .0277 "
100	100 + .0000 "	100 + .0000 "
120	120 − .0284 "	120 − .0339 "
140	140 − .0615 "	140 − .0721 "
160	160 − .0983 "	160 − .1134 "
180	180 − .1382 "	180 − .1571 "
200	200 − .1796 "	200 − .2018 "
220	220 − .2232 "	220 − .2478 "
240	240 − .2663 "	240 − .2932 "
260	260 − .3141 "	260 − .3420 "
280	280 − .3610 "	280 − .3897 "
300	300 − .4085 "	300 − .4377 "

The standard defined by Regnault is that of the constant-volume air-thermometer, with air at the density which it has when at the freezing-point under the pressure of 760 mm., or 29.9218 inches of mercury, and its indications are shown in comparison with the absolute scale by taking $\dfrac{\phi}{v} = 1$ in the second column of the preceding table. The greatest discrepance between 0° and 100° Cent. amounts to less than $\frac{1}{25}$ of a degree, and the discrepance at 300° Cent. is only four-tenths. The discrepancies of the constant - pressure air - thermometer, when the pressure is equal to the standard atmospheric pressure, or $\dfrac{p}{\Pi} = 1$, are somewhat greater, but still very small.

SECTION V. — *Physical Properties of Air Expressed according to the Absolute Thermodynamic Scale of Temperature.*

All the physical properties of a fluid of given constitution are completely fixed when its density and temperature are specified; and as it is these qualities which we can most conveniently regard as being immediately adjustable in any arbitrary manner, we shall generally consider them as the independent variables in formulæ expressing the pressure, the specific heats, and other properties of the particular fluid in any physical condition.

Let v be the volume (in cubic feet) of a unit mass (one pound) of the fluid, and t its absolute temperature; and let p be its pressure in the condition defined by these elements.

Let also e be the "mechanical energy"[*] of the fluid, reckoned from some assumed standard or zero state, that is, the sum of the mechanical value of the heat communicated to it, and of the work spent on it, to raise it from that zero state to the condition defined by (v, t); and let N and K be its specific heats with constant volume, and with constant pressure, respectively. Then, denoting, as before, the mechanical equivalent of the thermal unit by J, and the value of Carnot's function for the temperature t by μ, we have[†]

$$\frac{de}{dv} = \frac{J}{\mu}\frac{dp}{dt} - p \qquad \cdots \qquad (1)$$

$$N = \frac{1}{J}\frac{de}{dt} \qquad \cdots \qquad (2)$$

$$K = \frac{1}{J}\frac{de}{dt} + \frac{1}{J}\left(\frac{de}{dv} + p\right)\frac{\dfrac{dp}{dt}}{-\dfrac{dp}{dv}} \qquad \cdots \qquad (3)$$

From these we deduce by eliminating e,

$$K - N = \frac{1}{\mu}\frac{\left(\dfrac{dp}{dt}\right)^{2}}{-\dfrac{dp}{dv}} \qquad \cdots \qquad (4)$$

and

---

[*] Dynamical Theory of Heat, Part V., On the Quantities of Mechanical Energy Contained in a Fluid in Different States as to Temperature and Density, § 82. *Trans. Roy. Soc. Edin.*, Dec. 15, 1851.

[†] Ibid., §§ 89, 91.

$$\frac{dN}{dv} = \frac{d\left(\frac{1}{\mu}\frac{dp}{dt}\right)}{dt} - \frac{1}{J}\frac{dp}{dt} \quad \cdots \cdots \quad (5)$$

equations which express two general theorems regarding the specific heats of any fluid whatever, first published* in the *Transactions of the Royal Society of Edinburgh*, March, 1851. The former (4) is the extension of a theorem on the specific heats of gases originally given by Carnot,† while the latter (5) is inconsistent with one of his fundamental assumptions, and expresses, in fact, the opposed axiom of the Dynamical Theory. The use of the absolute thermodynamic system of thermometry proposed in Section IV., according to which the definition of temperature is

$$t = \frac{J}{\mu} \quad \cdots \cdots \cdots \quad (6)$$

simplifies these equations, and they become

$$JK - JN = t\frac{\left(\frac{dp}{dt}\right)^2}{-\frac{dp}{dv}} \quad \cdots \cdots \quad (7)$$

$$\frac{d(JN)}{dv} = t\frac{d^2p}{dt^2}. \quad \cdots \cdots \quad (8)$$

To compare with the absolute scale the indications of a thermometer in which the particular fluid (which may be any gas, or even liquid) referred to in the notation p, v, t is used as the thermometric substance, let p_0 and p_{100} denote the pressures which it has when at the freezing and boiling points respectively, and kept in constant volume, v; and let v_0 and v_{100} denote the volumes which it occupies under the same pressure, p, at those temperatures. Then if θ and ϑ denote its thermometric

---

* Dynamical Theory of Heat, Part V., On the Quantities of Mechanical Energy Contained in a Fluid in Different States as to Temperature and Density, §§ 47, 48.

† See "Account of Carnot's Theory," Appendix III., *Trans. Roy. Soc. Edin.*, April 30, 1849, p. 565.

indications when used as a constant-volume **and** as a constant-pressure thermometer respectively, we have

$$\theta = 100 \frac{p - p_0}{p_{100} - p_0} \quad \ldots \quad \ldots \quad (9)$$

$$\vartheta = 100 \frac{v - v_0}{v_{100} - v_0}. \quad \ldots \quad \ldots \quad (10)$$

Let also ε denote the "coefficient of increase of elasticity **with** temperature,"* and ϵ denote the coefficient of expansion **at** constant pressure, when **the** gas is in the **state** defined by (v, t); and let E and E denote **the mean** values of **the same** coefficients between $0°$ and $100°$ Cent. Then we have

$$\varepsilon = \frac{1}{p_0} \frac{dp}{dt} \quad \ldots \quad \ldots \quad (11)$$

$$\epsilon = \frac{\dfrac{dp}{dt}}{v_0 \times -\dfrac{dp}{dv}} \quad \ldots \quad \ldots \quad (12)$$

$$\mathrm{E} = \frac{p_{100} - p_0}{100 p_0} \quad \ldots \quad \ldots \quad (13)$$

$$E = \frac{v_{100} - v_0}{100 v_0}. \quad \ldots \quad \ldots \quad (14)$$

Lastly, the general expression for $\dfrac{J}{\mu}$, quoted in Section II. from our paper **of last** year, leads to the following expression for the cooling effect on **the** fluid when forced through a porous plug as in our air experiments :

$$\delta = \frac{1}{\mathrm{JK}} \left\{ \int_{\mathrm{V}}^{\mathrm{V}'} \left(t \frac{dp}{dt} - p \right) dv + (\mathrm{P}'\mathrm{V}' - \mathrm{PV}) \right\}, \quad (15)$$

(p, v), $(\mathrm{P}', \mathrm{V}')$, (P, V), as explained above, having reference to the fluid in different **states** of density, but always at the same temperature, t, as that **with which** it enters the plug.

---

* So called by **Mr. Rankine.** The same element is called by M. Regnault the coefficient of dilatation of a gas at constant volume.

From these equations it appears that if p be fully given in terms of v and absolute values for t for any fluid, the various properties denoted by

$$JK - JN, \frac{d(JN)}{dv}, \theta, \vartheta, \varepsilon, \epsilon, E, \mathcal{E}, \text{ and } \delta$$

may all be determined for it in every condition. Conversely, experimental investigations of these properties may be made to contribute, along with direct measurements of the pressure for various particular conditions of the fluid, towards completing the determination of the function which expresses this element in terms of v and t. But it must be remarked that even complete observations determining the pressure for every given state of the fluid could give no information as to the values of t on the absolute scale, although they might afford data enough for fully expressing p in terms of the volume and the temperature with reference to some particular substance used thermometrically. On the other hand, observations on the specific heats of the fluid, or on the thermal effects it experiences in escaping through narrow passages, may lead to a knowledge of the absolute temperature, t, of the fluid when in some known condition, or to the expression of p in terms of v, and absolute values of t; and accordingly the formulæ (7), (8), and (15) contain t explicitly, each of them, in fact, essentially involving Carnot's function. As for actual observations on the specific heats of air, none which have yet been published appear to do more than illustrate the theory, by confirming (as Mr. Joule's, and the more precise results more recently published by M. Regnault, do), within the limits of their accuracy, the value for the specific heat of air under constant pressure which we calculated* from the *ratio of the specific heats*, determined according to Laplace's theory by observations on the velocity of sound, and the *difference of the specific heats* determined by Carnot's theorem with the value of Carnot's function estimated from Mr. Joule's original experiments on the changes of temperature produced by the rarefaction and condensation of air† and established to a closer degree of accuracy by our preliminary experiments on expansion through a resisting solid.‡ It

---

* *Philosophical Transactions*, March. 1852, p. 82.

† *Royal Society Proceedings*, June 20, 1844; or *Phil. Mag.*, May, 1845.

‡ Ibid., December, 1850.

ought also to be remarked that the specific heats of air can only be applied to the evaluation of absolute temperature with a knowledge of the mechanical equivalent of the thermal unit; and, therefore, it is probable that, even when sufficiently accurate determinations of the specific heats are obtained, they may be useful rather for a correction or verification of the mechanical equivalent than for the thermometric object. On the other hand, a comparatively very rough approximation to JK, the mechanical value of the specific heat of a pound of the fluid, will be quite sufficient to render our experiments on the cooling effects available for expressing with much accuracy by means of the formula (15) a thermodynamic relation between absolute temperature and the mechanical properties of the fluid at two different temperatures.

In the notes to Mr. Joule's paper on the Air-Engine,[*] it was shown that if Mayer's hypothesis be true we must have approximately,

$$K = .2374 \text{ and } N = .1684,$$

because observations on the velocity of sound, with Laplace's theory, demonstrate that

$$k = 1.410,$$

within $\frac{1}{100}$ of its own value. Now the experiments at present communicated to the Royal Society prove a very remarkable approximation to the truth in that hypothesis (see above, Section I.), and we may therefore use these values as very close approximations to the specific heats of air. The experiments on the friction of fluids and solids made for determining the mechanical value of heat[†] give for J the value 1390; and we therefore have $JN = 234.1$ with sufficient accuracy for use in calculating small terms.

Now, according to Regnault, we have, for dry air at the freezing-point, in the latitude of Paris,

$$H = 26215 ;$$

and since the force of gravity at Paris, with reference to a foot as the unit of space and a second as the unit of time, is 32.1813,

---

it follows that the velocity of sound in dry air at 0° Cent. would be, according to Newton's unmodified theory,

$$\sqrt{26215 \times 32.1813} = 918.49 \; ;$$

or, in reality, according to Laplace's theory,

$$\sqrt{k} \; \sqrt{26215 \times 32.1813}.$$

But according to Bravais and Martins it is in reality

1090.5, which requires that $k = 1.4096$;

or, according to Moll and Van Beck,

1090.1, which requires that $k = 1.4086$.

The mean of these values of k is 1.4091.

[Note of January 5, 1882, by Sir W. Thomson.—That portion of this Second Part of our researches which was devoted to working out an empirical formula for the thermo-elastic properties of air, and the calculation of specific heats from it, is not reproduced here, because at the conclusion of Part IV. we have derived a better and simpler empirical formula from more comprehensive experimental data.]

ABSTRACT OF THE ABOVE PAPER "ON THE THERMAL EFFECTS OF FLUIDS IN MOTION."

No. II.

By J. P. JOULE, Esq., F.R.S., and PROFESSOR W. THOMSON, F.R.S. (*Proceedings of the Royal Society*, Vol. VII., p. 127.)

THE first experiments described in this paper show that the anomalies exhibited in the last table of experiments, in the paper preceding it,* are due to fluctuations of temperature in the issuing stream consequent on a change of the pressure with which the entering air is forced into the plug. It appears from these experiments that when a considerable alteration is suddenly made in the pressure of the entering stream, the issuing stream experiences remarkable successions of augmentations and diminutions of temperature, which are sometimes perceptible for half an hour after the pressure of the entering stream has ceased to vary.

Several series of experiments are next described in which air is forced (by means of a large pump and other apparatus described in the first paper) through a plug of cotton-wool, or unspun silk pressed together, at pressures varying in their excess above the atmospheric pressure from five or six up to fifty or sixty pounds on the square inch. By these it appears that the cooling effect which the air, as found in the author's previous experiments, always experiences in passing through the porous plug, varies proportionally to the excess of pressure of the air on entering the plug above that with which it is allowed to escape. Seven series of experiments, in each of which the air entered the plug at a temperature of about 16° Cent., gave a mean cooling effect of about 0°.0175 Cent. per pound on the square inch, or 0°.27 Cent. per atmosphere of dif-

* *Proc. Roy. Soc.*, and *Phil. Mag.*, September, 1853, p. 230.

ference of pressure. Experiments made at lower and at higher temperatures showed that the cooling effect is very sensibly less for high than for low temperatures, but have not yet led to sufficiently exact results at other temperatures than that stated (16° Cent.) to indicate the law according to which it varies with the temperature.

Experiments on carbonic acid at different temperatures are also described, which show that at about 16° Cent. this gas experiences $4\frac{1}{2}$ times as great a cooling effect as air. They agree well at all the different temperatures with a theoretical result, derived according to the general dynamical theory from empirical formulæ for the pressure of carbonic acid in terms of its temperature and density, which was kindly communicated by Mr. Rankine to the authors, having been investigated by him upon no other experimental data than those of Regnault on the expansion of gas by heat and its compressibility.

Experiments were also made on hydrogen gas, which, although not such as to lead to accurate determinations, appeared to indicate very decidedly a cooling effect* amounting to a small fraction, perhaps about $\frac{1}{10}$, of that which air would experience in the same circumstances.

The following theoretical deductions from these experiments are made :

I. The relations between the heat generated and the work spent in compressing carbonic acid, air, and hydrogen, are investigated from the experimental results. In each case the relation is nearly that of equivalence, but the heat developed exceeds the equivalent of the work spent by a very small amount for hydrogen, considerably more for air, and still more for carbonic acid. For slight compressions with the gases kept about the temperature 16°, this excess amounts to about $\frac{1}{17}$ of the whole heat emitted in the case of carbonic acid and $\frac{1}{120}$ in the case of air.

II. It is shown in the general dynamical theory that the air experiments taken in connection with Regnault's experimental results on the latent heat and pressure of saturated steam make it certain that the density of saturated steam increases very much more with the pressure than according to Boyle's and Gay-Lussac's gaseous laws, and numbers are given express-

[* *Later experiments showed this to be a mistake. See Part IV.*]

84

ing the theoretical densities of saturated steam, at different temperatures, which it is desired should be verified by direct experiments.

III. Carnot's function in the "Theory of the Motive Power of Heat" is shown to be very nearly equal to the mechanical equivalent of the thermal unit divided by the temperature from the zero of the air-thermometer (that is, temperature Centigrade with a number equal to the reciprocal of the coefficient of expansion added), and corrections, depending on the amount of the observed cooling effects in the new air experiments, and the deviations from the gaseous laws of expansion and compression determined by Regnault, are applied to give a more precise evaluation.

IV. An absolute scale of temperature—that is, a scale not founded on reference to any particular thermometric substance or to any special qualities of any class of bodies—is founded on the following definition :

If a physical system be subjected to cycles of perfectly reversible operations and be not allowed to take in or emit heat except in localities, at two fixed temperatures, these temperatures are proportional to the whole quantities of heat taken in or emitted at them respectively during a complete cycle of the operations.

The principles upon which the degree or unit of temperature is to be chosen, so as to make the difference of temperatures on the absolute scale agree with that on any other scale for a particular range of temperatures. If the difference of temperatures between the freezing- and the boiling-points of water be made 100° on the new scale, the absolute temperature of the freezing-point is shown to be about 273°.7; and it is demonstrated that the temperatures from the freezing-point on the new scale will agree very closely with Centigrade temperature by the standard air-thermometer; quite within the limits of the most accurate practical thermometry when the temperature is between 0° and 100° Cent., and very nearly, if not quite, within these limits for temperatures up to 300° Cent.

[V. An empirical formula for the pressure of air in terms of its density, and its temperature on the absolute scale, is investigated by using forms such as those first proposed and used by Mr. Rankine, and determining the constants so as to fulfil the conditions (1) of giving the observed cooling effects, (2) of agreeing with Regnault's experimental results on compressibility at a particular temperature.

A table of comparison of temperature by the air-thermometer under varied conditions of temperature and pressure with the absolute scale is deduced from this formula.]*

Expressions for the specific heat of any fluid in terms of the absolute temperature, the density, and the pressure, derived from the general dynamical theory, are worked out for the case of air according to the empirical formula; and tables of numerical results derived exclusively from these expressions and the ratio of the specific heats as determined by the theory of sound are given. These tables show the mechanical values of the specific heats of air at different constant pressures and at constant densities. Taking 1390 as the mechanical equivalent of the thermal unit as determined by Mr. Joule's experiment on the friction of fluids, the authors find, as the mean specific heat of air under constant pressure,

0.2390, from 0° to 100° Cent.
0.2384, from 0° to 300° Cent.

* [The section in brackets appears in the original contribution to the Proc. Roy. Soc. and in Thomson's Math. and Phys. Papers, but is omitted in Joule's Scientific Papers.]

On the Thermal Effects of Fluids in Motion.

Part IV.

By J. P. JOULE, LL.D., F.R.S., Etc., and PROFESSOR
WILLIAM THOMSON, A.M., LL.D., F.R.S., Etc.

(*Phil. Trans.*, 1862, p. 579.)

In the Second Part of these researches we have given the
results of our experiments on the difference between the temperatures of an elastic fluid on the high- and low-pressure sides
of a porous plug through which it was transmitted. The gases
employed were atmospheric air and carbonic acid. With the
former $0°.0176$ of cooling effect was observed for each pound
per square inch of difference of pressure, the temperature on
the high-pressure side being $17°.125$. With the latter gas
$0°.0833$ of cooling effect was produced per pound of difference
of pressure, the temperature on the high-pressure side being
$12°.844$.

It was also shown that in each of the above gases the difference of the temperatures on the opposite sides of the porous
plug is sensibly proportional to the difference of the pressures.

An attempt was also made to ascertain the cooling effect
when elastic fluids of high temperatures were employed; and
it was satisfactorily shown that in this case a considerable
diminution of the effect took place. Thus, in air at $91°.58$
the effect was only $0°.014$; and in carbonic acid at $91°.52$ it
was $0°.0474$.

In the experiments at high temperature there appeared to be
some grounds for suspecting that the apparent cooling effect
was too high; for the quantity of transmitted air was very considerable, and its temperature had possibly not arrived accurately at that of the bath by the time it reached the porous
plug.

The obvious way to get rid of all uncertainty on this head was to increase the length of the coil of pipes. Hence in the following experiments the total length of 2-inch copper pipe immersed in the bath was 60 feet, instead of 35, as in the former series. The volume of air transmitted in a given time was also considerably less. There could, therefore, be no doubt that the temperature of the air on its arrival at the plug was sensibly the same as that of the bath.

The nozzle employed in the former series of experiments was of boxwood—the space occupied by cotton-wool, or other porous material, being 2.72 inches long and an inch and a half in diameter. The boxwood was protected from the water of the bath by being enveloped by a tin can filled with cotton-wool. This was unquestionably in most respects the best arrangement for obtaining accurate results; but it was found necessary to make each experiment last one hour or more before we could confidently depend on the thermal effect. The oscillations of temperature which took place during the first part of the time were traced to various causes, one of the principal being the length of time which, on account of the large capacity for heat and the small conductivity of the boxwood nozzle, elapsed before the first large thermal effects consequent on the getting up of the pressure were dissipated. No doubt the results we arrived at were very accurate with the elastic fluids employed, viz., atmospheric air and carbonic acid; but we possessed an unlimited supply of the former and a supply of the latter equal to 120 cubic feet, which was sufficient to last for more than half an hour without being exhausted. In extending the inquiry to gases not so readily procured in large quantities, it was therefore desirable to use a porous plug of smaller dimensions enclosed in a nozzle of less capacity for heat, so as to arrive rapidly at the normal effect.

Various alterations of the apparatus were made in order to meet the requirements of our experiments. A small high-pressure engine of about one horse-power was placed in gear with a double-acting compressing air-pump, which had a cylinder four and a half inches in diameter, with a length of stroke of nine inches. The engine was able to work the piston of the pump sixty complete strokes in the minute. The quantity of air which it ought to have discharged at low pressure was therefore upwards of 16,000 cubic inches per minute. But

FREE EXPANSION OF GASES

PLATE III.—Fig. 1.

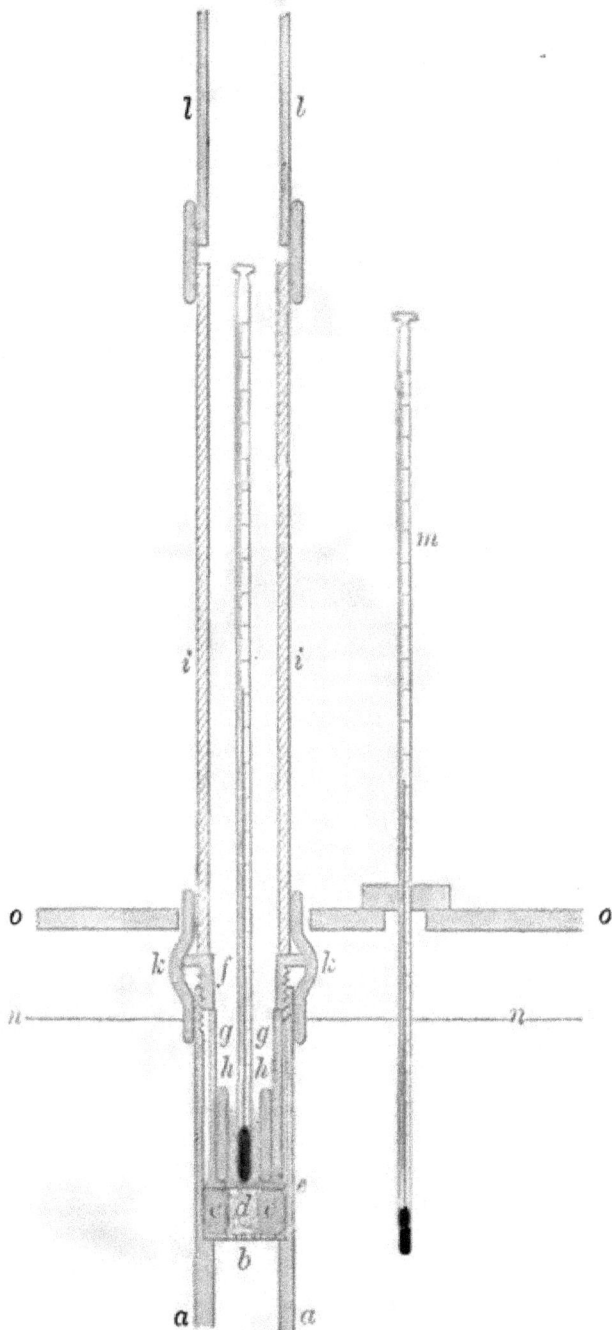

much loss, of course, occurred from leakage past the metallic piston, and in consequence of the necessary clearance at the top and bottom of the cylinder when the pressure increased by a few atmospheres; so that in practice we never pumped more than 8000 cubic inches per minute.

The nozzle we employed will be understood by inspecting Plate III., Fig. 1, where *a a* is the upright end of the coil of copper pipes. On the shoulder within the pipe a perforated metallic disk (*b*) rests. Over this is a short piece of india-rubber tube (*c c*) enclosing a silk plug (*d*), which is kept in a compressed state by the upper perforated metallic plate (*e*). This upper plate is pressed down with any required force by the operation of the screw *f* on the metallic tube *g g*. A tube of cork (*h h*) is placed within the metallic tube, in order to protect the bulb of the thermometer from the effects of a too rapid conduction of heat from the bath. Cotton-wool is loosely packed round the bulb, so as to distribute the flowing air as evenly as possible. The glass tube (*i i*) is attached to the nozzle by means of a strong piece of india-rubber tubing [*k k*], and through it the indications of the thermometer are read. The top of the glass tube is attached to the metallic tube *l l*, for the purpose of conveying the gas to the meter.

The thermometer (*m*) for registering the temperature of the bath is placed with its bulb near the nozzle. The level of the water is shown by *n n*; and *o o* represents the wooden cover of the bath. When a high temperature was employed it was maintained by introducing steam into the bath by means of a pipe led from the boiler. The water of the bath was in every case constantly and thoroughly stirred, especially when high temperatures were used.

The general disposition of the apparatus will be understood from Fig. 2, in which A represents the boiler, B the steam-engine geared to the condensing air-pump C. From this pump the compressed air passes through a train of pipes 60 feet long and 2 inches in diameter, and then enters the coil of pipes in the bath D. Thence, after issuing from the porous plug, it passes through the gasometer E, and ultimately arrives again at the pump C. This complete circulation is of great importance, inasmuch as it permits the gas which has been collected in the meter to be used for a much longer period than would otherwise have been possible. A glass vessel full of

Fig. 2.

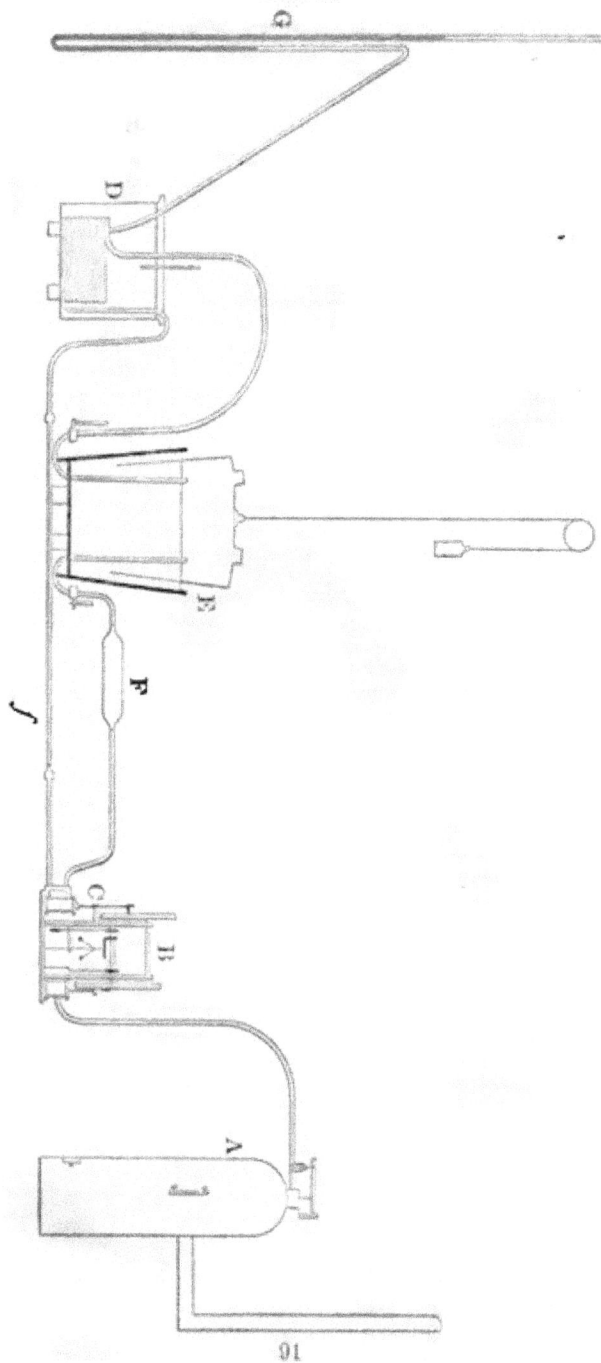

chloride of calcium is also placed in the pipe at f. A small tube leading from the coil is carried to the shorter leg of the glass siphon gauge G, of which the longer leg is 17 feet and the shorter 12 feet long.

The thermometers employed were all carefully calibrated, and had about ten divisions to the degree Centigrade. We took the precaution of verifying the air- and bath-thermometers from time to time, especially when high temperatures were used, in which latter case a comparison between the thermometers at high temperature was made immediately after each experiment.

Atmospheric Air. (See Table I.)

In the experiments described in the present paper the air was not deprived of its carbonic acid. It was simply dried by transmitting it in the first place, before it entered the pump, through a cylinder 18 inches long and 12 inches in diameter filled with chloride of calcium, and afterwards, in its compressed state, through a pipe 12 feet long and 2 inches in diameter filled with the same substance. The experiments were principally carried on in the winter season; so that the chloride kept dry for a long time. From its condition after some weeks' use, it was evident that the water was removed, almost as much as chloride of calcium can remove it, after the air had traversed three inches of the chloride contained by the first vessel.

Oxygen Gas. (See Table II.)

This elastic fluid was procured by cautiously heating chlorate of potash mixed with a small quantity of peroxide of manganese. In its way to the meter it passed through a tube containing caustic potash, in order to deprive it of any carbonic acid it might contain. The same drying apparatus was employed as in the case of atmospheric air.

Nitrogen Gas. (See Table III.)

In preparing this gas the meter was first filled with air, and then a long, shallow tin vessel was floated under it, containing sticks of phosphorus so disposed as to burn in succession. Some hours were allowed to elapse after the combustion had terminated, in order to allow of the deposition of phosphoric acid formed.

FREE EXPANSION OF GASES

TABLE 1.

No. of Experiment	Cubical Inches of Air Transmitted per Minute	Pressure over that of the Atmosphere, in Inches of Mercury	Temperature of the Bath	Thermal Effect	Correction on Account of Conduction of Heat	Corrected Thermal Effect	Thermal Effect Reduced to the Pressure of 100 Inches of Mercury	Time Occupied by Experiment, in Minutes	Number of Observations Comprised in Each Mean	Extreme Range of the Temperature of the Bath	Extreme Range of the Temperature of the Air	Extreme Range of the Pressure
1	3000	83.96	4.499	−0.711	−0.041	−0.752	−0.900	14	5	0.020	0.015	2.25
2	3600	136.14	6.112	−1.11	−0.058	−1.168	−0.858	24	7	0.017	0.035	1.7
3	2600	136.59	6.082	−1.307	−0.094	−1.401	−0.895	18	15	0.065	0.065	8.0
4	1750	130.58	7.471	−1.137	−0.122	−1.259	−0.902	18	15	0.009	0.10	2.0
5	2250	153.9	7.640	−1.231	−0.103	−1.334	−0.807	12.5	10	0.006	0.135	7.0
6	2300	150.3	8.546	−1.252	−0.102	−1.354	−0.850	24	20	0.008	0.198	7.0
7	3900	165.73	8.2	−1.329	−0.121	−1.450	−0.875	26	15	0.017	0.105	4.8
8	2000	129.73	8.72	−1.019	−0.127	−1.146	−0.883	12	9	0.084	0.085	2.9
9	1300	128.91	24.92	−0.983	−0.037	−1.020	−0.791	8	8	0.028	0.06	7.0
10	5000	122.8	27.81	−0.874	−0.036	−0.910	−0.741	6	4	0.015	0.064	3.6
11	4600	123.5	42.64	−0.947	−0.036	−0.983	−0.796	25	30	0.127	0.122	5.3
12	5000	157	43.54	−0.943	−0.037	−0.980	−0.764	28	30	0.058	0.09	8.0
13	4800	127.5	47.92	−0.937	−0.037	−0.974	−0.715	24	20	0.14	0.02	8.0
14	5000	147	47.96	−0.939	−0.049	−0.988	−0.692	20	10	0.65	0.09	2.0
15	3700	146	49.96	−0.890	−0.028	−0.918	−0.608	19	3	0.18	0.26	7.0
16	5000	146	53.355	−0.870	−0.028	−0.898	−0.616	12	10	0.112	0.08	8.6
17	5600	112.43	64.9	−0.469	−0.029	−0.502	−0.446	4	10	0.022	0.23	7.0
18	5700	147	89.901	−0.821	−0.091	−0.912	−0.620	6	20	0.202	0.23	3.5
19	1700	153.16	90.353	−0.756	−0.083	−0.839	−0.547	8	16	0.078	0.273	0
20	1700	136.5	92.486	−0.674	−0.040	−0.714	−0.504	24	20	0.236	0.19	0
21	3130	146	92.603	−0.700	−0.036	−0.736	−0.504	20	20	0.112	0.255	0
22	4600	138.5	93.78	−0.722	−0.029	−0.751	−0.474	30	16	0.112	0.115	0

TABLE II.

1	2	3	4	5	6	7	8	9	10	11	12	13	14	15
No. of Experiment	Cubical Inches per Minute Transmitted of Elastic Fluid	Composition of the Elastic Fluid	Pressure over that of the Atmosphere, in Inches of Mercury	Temperature of the Bath	Thermal Effect	Correction on Account of Conduction of Heat	Corrected Thermal Effect	Thermal Effect Reduced to the Pressure of 100 Inches of Mercury	Ditto Calculated for Pure Oxygen	Time Occupied by Experiment, in Minutes	Number of Observations Comprised in Each Mean	Extreme Range of the Temperature of the Bath	Extreme Range of the Temperature of the Elastic Fluid	Extreme Range of the Pressure
1	2000	N { 5.095 } O { 94.905 }	158	88.682	−1.547	−0.145	−1.692	−1.061	−1.025	9	10	0.005	0.35	2.0
2	2000	N { 51.62 } O { 45.38 }	161	88.75	−1.373	−0.129	−1.502	−0.928	−1.074	11	10	0.015	0.046	0.9
3	1700	N { 3.61 } O { 96.37 }	151	89.486	−1.059	−0.118	−1.187	−0.786	−0.890	11	10	0.43	0.43	6.2
4	3150	N { 22.37 } O { 77.63 }	159.77	90.8	−0.840	−0.050	−0.890	−0.557	−0.578	12	10	0.326	0.336	8.0
5	3150	N { 51.03 } O { 48.97 }	154.1	92.742	−0.734	−0.043	−0.777	−0.501	−0.527	12	10	0.18	0.19	4.0
6	4500	N { 4 } O { 96 }	152	93.473	−0.795	−0.033	−0.828	−0.544	−0.570	11	8	0.135	0.158	9

TABLE III.

1	2	3	4	5	6	7	8	9	10	11	12	13	14	15
No. of Experiment	Cubical Inches per Minute Transmitted of Elastic Fluid	Composition of the Elastic Fluid	Pressure over that of the Atmosphere, in Inches of Mercury	Temperature of the Bath	Thermal Effect	Correction on Account of Conduction of Heat	Corrected Thermal Effect	Thermal Effect Reduced to the Pressure of 100 Inches of Mercury	Ditto Calculated for Pure Nitrogen	Time Occupied by Experiment, in Minutes	Number of Observations Comprised in Each Mean	Extreme Range of the Temperature of the Bath	Extreme Range of the Temperature of the Elastic Fluid	Extreme Range of the Pressure
1	2050	N { 79.12 } O { 8.65 }	163.38	7.204	−1.448	−0.133	−1.581	−0.965	−1.031	7	8	0.008	0.25	6.0
2	2500	N { 81.32 } O { 17.56 }	162.65	91.415	−0.857	−0.064	−0.921	−0.567	−0.526	13	10	0.036	0.48	4.5
3	2500	N { 12.1 }	164.61	91.965	−0.869	−0.065	−0.934	−0.567	−0.891	12	9	0.337	0.378	3.0

TABLE IV.

1. No. of Experiment	2. Cubical Inches of Elastic Fluid Transmitted per Minute	3. Composition of the Elastic Fluid	4. Pressure over that of the Atmosphere in Inches of Mercury	5. Temperature of the Bath	6. Thermal Effect	7. Correction on Account of Conduction of Heat	8. Corrected Thermal Effect	9. Thermal Effect Referred to the Pressure of 100 Inches of Mercury	10. Ditto Calculated for Pure Carbonic Acid, calling the Specific Heat for Equal Vol. 1.29	11. Time Occupied by Experiment, in Minutes	12. Number of Observations Combined in Each Mean	13. Extreme Range of the Temperature of the Bath	14. Extreme Range of the Temperature of the Elastic Fluid	15. Extreme Range of the Pressure
1	2480	Air 66.42 / CO₂ 31.58 / Air 89.16	163.7	0.322	−2.099	−0.130	−2.229	−1.765	−3.166	12	16	0	0.16	0.2
2	2330	Air 10.84 / CO₂ 3.52 / CO₂ 96.48	163.82	7.360	−1.621	−0.125	−1.746	−1.173	−2.990	14	19	0.004	0.282	9.2
3	3100	Air 62.5 / CO₂ 37.5 / Air 88.13	164.07	7.384	−6.719	−0.299	−7.018	−4.277	−4.367	6.5	6	0.008	0.021	1.4
4	2500	CO₂ 11.87 / Air 97.46 / CO₂ 2.54	162.925	7.497	−2.839	−0.191	−3.030	−1.890	−3.952	8	8	0.007	0.11	5.6
5	2300	Air 4.0 / H 5.286 / Air 90.714	158.08	7.433	−1.682	−0.132	−1.814	−1.147	−2.648	10	10	0.005	0.307	5.2
6	2250	CO₂ 4.23 / Air 46.47 / CO₂ 49.3	163.52	7.608	−1.407	−0.116	−1.523	−0.931	−2.753	8	8	0.007	0.064	2.0
7	3380	H 7.09 / Air 67.95 / CO₂ 25.86	161.97	7.960	−6.131	−0.262	−6.393	−3.947	−4.215	6	8	0	0.18	4.8
8	3600	Air 2.11 / H 97.89 / Air 56.78	163.72	8.029	−2.189	−0.117	−2.306	−1.500	−2.631	6	6	0	0.19	1.6
9	1900	CO₂ 49.22 / Air 77.77 / CO₂ 22.28	97.56	8.296	−0.543	−0.063	−0.606	−0.622	−1.940	15	15	0	0.146	5.4
10	2925	Air 0.83 / CO₂ 99.17 / Air 67.7	167.25	93.525	−3.418	−0.160	−3.578	−2.139	−2.164	10	10	0.012	0.49	4.0
11	2925	CO₂ 32.3 / Air 87.77 / CO₂ 12.23	167.4	91.26	−1.746	−0.099	−1.845	−1.102	−1.674	30	20	0.382	0.49	11.0
12	2925	Air 1.83 / CO₂ 98.17 / Air 1.66	146.83	91.642	−1.292	−0.077	−1.369	−0.938	−2.053	9	6	0.292	0.248	3.5
13	5500	CO₂ 98.34	146	54.0	−4.184	−0.104	−4.288	−2.937	−2.951	24	16	0.045	0.46	0
14	5300	Air	147	49.703	−1.832	−0.059	−1.891	−1.296	−2.225	24	16	0.24	0.17	0
15	5500	CO₂	145	49.764	−1.250	−0.032	−1.282	−0.884	−2.025	20	16	0.025	0.11	0
16	5100	Air	127.5	35.601	−4.196	−0.112	−4.308	−3.371	−3.407	18	15	0.01	0.095	0
17	6000	CO₂	151	97.653	−3.11	−0.084	−3.194	−2.115	−2.135	20	16	0.03	0.272	0

TABLE V.

No. of Experiment	Cubical Inches of Elastic Fluid Trans-mitted per Minute	Compression of the Elastic Fluid	Pressure over that of the Atmosphere, in Inches of Mercury	Temperature of the Bath	Thermal Effect	Correction on Account of Conduction of Heat	Corrected Thermal Effect	Thermal Effect Reduced to the Pressure of 100 Inches of Mercury	Data Calcu-lated for Pure Hydrogen	Time Occupied by Experi-ment, in Minutes	Number of Observations Comprised in Each Mean	Extreme Range of Temperature of the Bath	Extreme Range of Temperature of the Elastic Fluid	Extreme Range of the Pressure
1	2	3	4	5	6	7	8	9	10	11	12	13	14	15
1	3000	17.965 Air / 82.865 H / 25.16 Air	64.1	6.34	0.144	0.009	0.153	0.259	−0.104	3	3	0	0	0
2	3000	21.84 Air / 4.866 H	99.96	6.335	0.564	0.035	0.599	0.699	+0.226	10	4	0	0.15	1.5
3	3000	95.134 H / 78.295 Air	49.91	6.132	0.633	0.092	0.343	0.070	+0.118	12	6	0.002	0.06	1.2
4	2960	21.705 H / 0.2	99.657	5.808	0.536	0.014	0.362	0.571	+0.525	34	12	0.003	0.11	1.5
5	2960	1.798 Air / 98.802 H	86.885	7.244	0.041	0.003	0.044	0.05	+0.143	27	10	0.004	0.088	1.75
6	3000	1.795 Air / 93.205 H	79.84	7.652	0.043	0.004	0.046	0.028	+0.075	25	8	0.008	0.023	2.85
7	2950	67.75 Air / 32.25 H	74.08	6.654	0.054	0.004	0.058	0.028	+0.126	17	10	0.016	0.11	6.6
8	2950	4.07 Air / 95.93 H	100.47	6.717	0.571	0.040	0.611	0.466	+0.383	12	6	0.021	0.07	2.6
9	3000	68.29 H / 41.71 Air	100.72	6.781	0.639	0.002	0.041	0.041	+0.08	10	10	0.012	0.058	2.6
10	2100	91.91 H / 16.19 Air	114.02	6.846	0.304	0.035	0.539	0.355	+0.317	8.5	8	0.011	0.07	3.6
11	1900	27.66 Air / 1.375 H	153.67	7.400	1.002	0.099	1.101	0.721	+0.004	13	8	0	0.225	9.0
12	1700	95.625 H / 6.48 Air	138.55	7.474	1.032	0.11	1.142	0.825	+0.014	14	6	0.041	0.053	8.2
13	3600	53.752 Air / 5.035 H	87.74	89.66	0.151	0.005	0.189	0.215	+0.245	18	8	0.08	0.17	4.6
14	3000	94.957 H / 2.23 Air	91.72	92.951	0.061	0.005	0.056	0.034	+0.132	29	10	0.157	0.07	3.2
15	3600	97.441 H / 4.13 Air	72.99	90.354	0.062	0.004	0.057	0.104	+0.136	42	15	0.14	0.11	4.45
16	3600	2.59 / 425.87	85.15	89.292	0.111	0.004	0.128	0.139	+0.129	15.5	10	0.472	0.44	3.2
17	2900		101.72	89.858	0.073	0.004	0.057	0.073	+0.088			0.49	0.486	6.2

Carbonic Acid. (See Table IV.)

This gas was formed by adding sulphuric acid to a solution of carbonate of soda. It was dried in the same manner as all the other gases.

Hydrogen. (See Table V.)

Our method of procuring this elastic fluid was to pour sulphuric acid, prepared from sulphur, into a carboy nearly filled with water and containing fragments of sheet zinc. The gas was passed through a tube filled with rags steeped in a solution of sulphate of copper, and then through a tube filled with sticks of caustic potash. The rags became speedily browned, and we therefore adopted the plan of pouring a small quantity of solution of sulphate of copper from time to time into the carboy itself. This succeeded perfectly; the rags retained their blue color, and the gas was rendered perfectly inodorous, while at the same time its evolution became much more free and regular.

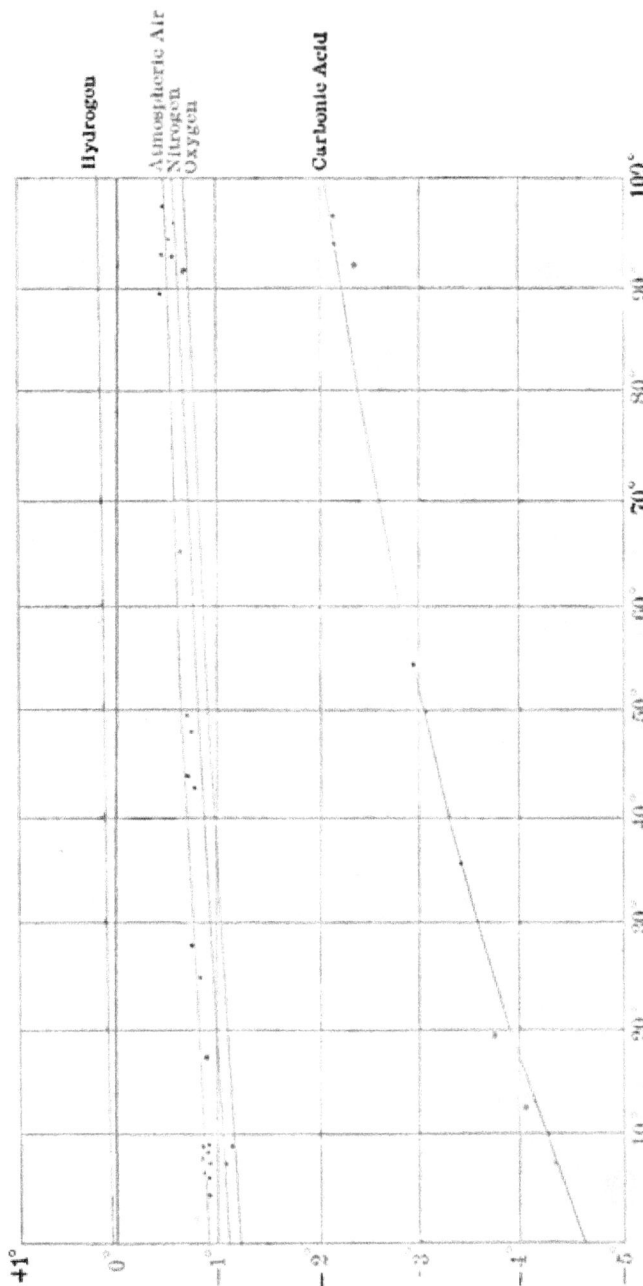

Chart of Thermal Effects for pressure of 100 inches mercury.

The asterisks show the results of the previous paper for Air and Carbonic Acid

Remarks on the Tables.

The correction of conduction of heat through the plug, inserted in column 6 of Table I., and in column 7 of the rest of the Tables, was obtained from data furnished by experiments in which the difference between the temperature of the bath and the air was purposely made very great. It was considered as directly proportional to the difference of temperature, and inversely to the quantity of elastic fluid transmitted in a given time.

The 10th column of Tables II., III., IV., and V. is calculated on the hypothesis that, in mixtures with other gases, atmospheric air retains its thermal qualities without change. This hypothesis is almost certainly incorrect, since it is reasonable to expect that the effect of mixture on the physical character is experienced by each of the constituent gases. The column is given as one method of showing the effect of mixture.

Effect of Mixture on the Constituent Gases. — Although the experiments on nitrogen given in Table III. are not so numerous as might be desired, we may infer from them, and the results in Table II., that common air and all other mixtures of nitrogen and oxygen behave more like a perfect gas, *i. e.*, give less cooling effect than either one or the other gas alone. We might expect the mixture to be something intermediate between the two. But this does not appear to be the case. The two are very nearly equal in their deviations from the condition of a perfect gas. Nitrogen deviates less than oxygen, but oxygen mixed with nitrogen differs less than nitrogen !

In the case of carbonic acid, which at low temperatures (7°) deviates five times as much as atmospheric air, we might expect that a mixture of CO_2 and air would deviate more than air and less than CO_2. This is the case (see Table IV.). Further, we might expect the two to contribute each its own proportion of cooling effect according to its own amount, and its specific heat volume for volume. But do the mixtures exhibit such a result ? No ! See column 10, Table IV., in which also note, under experiments 8 and 9, the great diminuation produced by the admixture of hydrogen.

If, instead of attributing to air and carbonic acid moments in proportion to their specific heats, or 1 : 1.39, as we have done in column 10, we use $1 : .7$, we obtain more consistent results.

Let δ denote the cooling effect experienced by air per 100 inches mercury, δ' that by carbonic acid, and Δ that by a mixture of volume V of air, and V' of carbonic acid; then we may take

$$\Delta = \frac{mV\delta + m'V'\delta'}{mV + m'V'}$$

to represent the cooling effect for the mixture, where m and m' are numbers which we may call the moments (or importances) of the two in determining the cooling effect for the mixture. The ratio of m to m' is the proper result of each experiment on a mixture, if we knew with perfect accuracy the cooling effect for each gas with none of the other mixed. Now, for common air we have direct experiments (Table I.), and know the cooling effect for it better than from any inferences from mixtures. But for pure CO_2 we know the effect for the most part only inferentially. Hence, having tried making $m : m' :: 1 : 1.39$ without obtaining consistent results, we tried other proportions, and, after various attempts, found that $m : m' :: 1 : .7$, for all temperatures and pressures within the limits of our experiments, gives results as consistent with one another as the probable errors of the experiments justify us in expecting. Thus, using the formula

$$\Delta = \frac{V\delta + V'\delta' \times .7}{V + V' \times .7},$$

we have, for calculating the effect of CO_2 from any experiment on a mixture, the following formula,

$$\delta' = \frac{(V + V' \times .7)\Delta - V\delta}{V' \times .7}.$$

Hence, using the numbers in columns 3 and 9 of Table IV., which relate to mixtures of air and carbonic acid alone, we find :

TABLE VI.

No. of Experiment	Proportions of Mixtures		Temperature of Bath	Thermal Effect for Air	Deduced Thermal Effect for Pure CO_2
	Air	CO_2			
1	68.42	31.58	7.36	−.88	−4.51
2	89.16	10.84	7.36	−.88	−4.61
3	3.52	96.48	7.38	−.88	−4.46
4	62.5	37.5	7.41	−.88	−4.19
5	88.13	11.87	7.43	−.88	−3.98
6	97.46	2.54	7.61	−.88	−3.89
16	1.83	98.17	35.6	−.75	−3.44
14	67.7	32.3	49.7	−.70	−3.04
15	87.77	12.23	49.76	−.70	−2.77
13	0.83	99.17	54	−.66	−2.96
10	2.11	97.89	93.52	−.51	−2.19
11	56.78	43.22	91.26	−.51	−2.21
12	77.77	22.23	91.64	−.51	−3.08
17	1.66	98.34	97.55	−.49	−2.16
1	2		3	4	5

The agreement for each set of results at temperatures nearly agreeing (with one exception, No. 12) shows that the assumption $m : m :: 1 : .7$ cannot be far wrong within our limits of temperature.

[*For continuation, see Thomson's Papers, Vol. I., p. 427, and Joule's Scientific Papers, Vol. II., p. 357. An empirical formula for the elasticity of gases is deduced.*]

On the Thermal Effects of Fluids in Motion.

Part IV.

By J. P. JOULE, LL.D., F.R.S., and PROFESSOR W. THOMSON, A.M., F.R.S.

(Abstract. *Proceedings of the Royal Society*, Vol. XII., p. 202.)

A BRIEF notice of some of the experiments contained in this paper has already appeared in the *Proceedings*. Their object was to ascertain with accuracy the lowering of temperature, in atmospheric air and other gases, which takes place on passing them through a porous plug from a state of high to one of low pressure. Various pressures were employed, with the result (indicated by the authors in their Part II.) that the thermal effect is approximately proportional to the difference of pressure on the two sides of the plug. The experiments were also tried at various temperatures, ranging from 5° to 98° Cent.; and have shown that the thermal effect, if one of cooling, is approximately proportional to the inverse square of the absolute temperature. Thus, for example, the refrigeration at the freezing temperature is about twice that at 100° Cent. In the case of hydrogen, the reverse phenomenon of a rise of temperature on passing through the plug was observed, the rise being doubled in quantity when the temperature of the gas was raised to 100°. This result is conformable with the experiments of Regnault, who found that hydrogen, unlike other gases, has its elasticity increased more rapidly than in the inverse ratio of the volume. The authors have also made numerous experiments with mixtures of gases, the remarkable result being that the thermal effect (cooling) of the compound gas is less than it would be if the gases after mixture retained in integrity the physical characters they possessed while in a pure state.

BOOKS OF REFERENCE

Bertrand, *Thermodynamique*, p. 66.
Maxwell, *Theory of Heat*, 7th Edition, p. 209.
Tait, *Heat*, p. 334.
Winkelmann, *Handbuch der Physik*, pp. 466–469, Vol. II., 2
Thomson's *Mathematical and Physical Papers*, Vol. I., p. 333.
v. der Waals, *Continuity of the Liquid and Gaseous States*.

ARTICLES

Natanson, *Wied. Ann.*, **31**, 502, 1887.
Lemoine, *Journal de Physique*, (2), **9**, 99, 1890.
Schiller, *Wied. Ann.*, **40**, 149, 1890.
Rose-Innes, *Proc. Phys. Soc. London*, December, 1897.
Bouty, *Journal de Physique*, (2), **8**, 20, 1889.
Regnault, *Mém. de Paris*, **37**, 579–959, 1868.
Clausius, *Wied. Ann.*, **9**, 337–357, 1880.
A Numerical Evaluation of the Absolute Scale of Temperature—R. A.
 Lehfeldt, *Phil. Mag.*, April (No. 275), 1898.

INDEX

INDEX

106

THE END